Numerical Solution of Initial-Value Problems in Differential-Algebraic Equations

SIAM's Classics in Applied Mathematics series consists of books that were previously allowed to go out of print. These books are republished by SIAM as a professional service because they continue to be important resources for mathematical scientists.

Editor-in-Chief
Gene H. Golub, *Stanford University*

Editorial Board
Richard A. Brualdi, *University of Wisconsin-Madison*
Herbert B. Keller, *California Institute of Technology*
Ingram Olkin, *Stanford University*
Robert E. O'Malley, Jr., *University of Washington*

Classics in Applied Mathematics

C. C. Lin and L. A. Segel, *Mathematics Applied to Deterministic Problems in the Natural Sciences*

Johan G. F. Belinfante and Bernard Kolman, *A Survey of Lie Groups and Lie Algebras with Applications and Computational Methods*

James M. Ortega, *Numerical Analysis: A Second Course*

Anthony V. Fiacco and Garth P. McCormick, *Nonlinear Programming: Sequential Unconstrained Minimization Techniques*

F. H. Clarke, *Optimization and Nonsmooth Analysis*

George F. Carrier and Carl E. Pearson, *Ordinary Differential Equations*

Leo Breiman, *Probability*

R. Bellman and G. M. Wing, *An Introduction to Invariant Imbedding*

Abraham Berman and Robert J. Plemmons, *Nonnegative Matrices in the Mathematical Sciences*

Olvi L. Mangasarian, *Nonlinear Programming*

*Carl Friedrich Gauss, *Theory of the Combination of Observations Least Subject to Errors: Part One, Part Two, Supplement*. Translated by G. W. Stewart

Richard Bellman, *Introduction to Matrix Analysis*

U. M. Ascher, R. M. M. Mattheij, and R. D. Russell, *Numerical Solution of Boundary Value Problems for Ordinary Differential Equations*

K. E. Brenan, S. L. Campbell, and L. R. Petzold, *Numerical Solution of Initial-Value Problems in Differential-Algebraic Equations*

Numerical Solution of Initial-Value Problems in Differential-Algebraic Equations

K. E. Brenan
The Aerospace Corporation
Los Angeles, California

S. L. Campbell
North Carolina State University
Raleigh, North Carolina

L. R. Petzold
University of Minnesota
Minneapolis, Minnesota

siam.

Society for Industrial and Applied Mathematics
Philadelphia

Copyright © 1996 by the Society for Industrial and Applied Mathematics.

This SIAM edition is an unabridged, corrected republication of the work first published by North-Holland, New York, 1989.

10 9 8 7 6 5 4 3 2 1

All rights reserved. Printed in the United States of America. No part of this book may be reproduced, stored, or transmitted in any manner without the written permission of the publisher. For information, write to the Society for Industrial and Applied Mathematics, 3600 University City Science Center, Philadelphia, PA 19104-2688.

Library of Congress Cataloging-in-Publication Data

Brenan, Kathryn Eleda, 1954–
   Numerical solution of initial-value problems in differential
-algebraic equations / K.E. Brenan, S. L. Campbell, L. R. Petzold.
      p. cm. -- (Classics in applied mathematics ; 14)
   "An unabridged, corrected republication of the work first
published by North-Holland, New York, 1989"--T.p. verso.
   Includes bibliographical references (p. -) and index.
   ISBN 0-89871-353-6 (pbk.)
   1. Initial value problems--Numerical solutions.   I. Campbell, S.
L (Stephen La Vern) II. Petzold, Linda Ruth. III. Title.
IV. Series.
QA378.B73   1996
512'.56--dc20   95-46482

siam. is a registered trademark.

Contents

Preface to the Classics Edition vii

Preface ix

1 **Introduction** 1
   1.1 Why DAE's? 1
   1.2 Basic Types of DAE's 3
   1.3 Applications 4
      1.3.1 Constrained Variational Problems 4
      1.3.2 Network Modeling 6
      1.3.3 Model Reduction and Singular Perturbations 9
      1.3.4 Discretization of PDE's 10
   1.4 Overview 13

2 **Theory of DAE's** 15
   2.1 Introduction 15
   2.2 Solvability and the Index 16
   2.3 Linear Constant Coefficient DAE's 18
   2.4 Linear Time Varying DAE's 22
      2.4.1 Solvability and the Index 22
      2.4.2 Structural Forms 25
      2.4.3 General Structure Theory 29
   2.5 Nonlinear Systems 32
      2.5.1 Solvability and the Index 32
      2.5.2 Structural Forms 34
      2.5.3 Index Reduction and Constraint Stabilization 36

3 **Multistep Methods** 41
   3.1 Introduction 41
   3.2 BDF Convergence 47
      3.2.1 Semi-Explicit Index One Systems 47
      3.2.2 Fully-Implicit Index One Systems 47
      3.2.3 Semi-Explicit Index Two Systems 54
      3.2.4 Index Three Systems of Hessenberg Form 56
   3.3 BDF Methods, DAE's, and Stiff Problems 58
   3.4 General Linear Multistep Methods 63

4 One-Step Methods — 75
- 4.1 Introduction — 75
- 4.2 Linear Constant Coefficient Systems — 78
- 4.3 Nonlinear Index One Systems — 86
  - 4.3.1 Semi-Explicit Index One Systems — 86
  - 4.3.2 Fully-Implicit Index One Systems — 88
- 4.4 Semi-Explicit Nonlinear Index Two Systems — 101
- 4.5 Order Reduction and Stiffness — 106
- 4.6 Extrapolation Methods — 108

5 Software for DAE's — 115
- 5.1 Introduction — 115
- 5.2 Algorithms and Strategies in DASSL — 117
  - 5.2.1 Basic Formulas — 117
  - 5.2.2 Nonlinear System Solution — 121
  - 5.2.3 Stepsize and Order Selection Strategies — 125
- 5.3 Obtaining Numerical Solutions — 129
  - 5.3.1 Getting Started with DASSL — 129
  - 5.3.2 Troubleshooting DASSL — 132
  - 5.3.3 Extensions to DASSL — 135
  - 5.3.4 Determining Consistent Initial Conditions — 138
- 5.4 Solving Higher Index Systems — 140
  - 5.4.1 Alternate Forms for DAE Systems — 140
  - 5.4.2 Solving Higher Index Systems Directly — 144

6 Applications — 149
- 6.1 Introduction — 149
- 6.2 Systems of Rigid Bodies — 150
- 6.3 Trajectory Prescribed Path Control — 157
- 6.4 Electrical Networks — 170
- 6.5 DAE's Arising from the Method of Lines — 177

Bibliography — 189

7 The DAE Home Page — 209
- 7.1 Introduction — 209
- 7.2 Theoretical Advances — 210
- 7.3 Numerical Analysis Advances — 214
- 7.4 Software Tools for DAE-Solving Environments — 216
- 7.5 The DASSL Family — 220

Bibliography — 239

Index — 253

Preface to the Classics Edition

In the last decade the use of differential-algebraic equations (DAE's) has become standard modeling practice in many applications, such as constrained mechanics and chemical process simulation. The advantages of using implicit, often computer-generated, models for dynamical processes are encouraging the use of DAE's in new areas.

The information in the 1989 edition of this book is still timely. We believe that those six chapters provide a good introduction to DAE's, to some of the mathematical and numerical difficulties in working with them, and to numerical techniques that are available for their numerical solution. The BDF-based code DASSL, carefully discussed in the first edition, still is arguably the best general-purpose index one DAE integrator.

This edition includes a new chapter, Chapter 7. A discussion of everything that has happened concerning DAE's and their applications would fill several more volumes. Chapter 7 has three more modest goals. First, it points out where improvements in some of the theorems concerning numerical methods, such as improved order estimates, have occurred. Second, there has been significant progress in the analytical understanding of higher index nonlinear DAE's. This progress is briefly outlined. Finally, some of the newly available software is described. In particular, several new extensions of DASSL are discussed, as well as some of the implicit Runge-Kutta-based codes. This combination of general information and a large bibliography of additional references that point to the relevant literature led us to title this new chapter "The DAE Home Page."

The colleagues and fellow scientists whose ideas have influenced us over the last decade are too numerous to mention. Many, but not all, appear in the bibliographies. Some who had an early impact on us are mentioned in the preface to the first edition. Special thanks are due to the Aerospace Corporation, the Air Force Office of Scientific Research, the Army Research Office, the Department of Energy, and the National Science Foundation for their funding and support, which made much of our research possible. We would like to thank Alan Hindmarsh, Peter Brown, and Laurent Jay for their careful reading of and suggestions for Chapter 7. Finally, this edition would not exist without the support and encouragement of Gene Golub, the series editor, and Vickie Kearn of SIAM.

September 15, 1995 *K. E. Brenan*
S. L. Campbell
L. R. Petzold

Preface

Differential-algebraic equations (DAE's) arise naturally in many applications, but present numerical and analytical difficulties which do not occur with ordinary differential equations. The numerical solution of these types of systems has been the subject of intense research activity in the past few years. A great deal of progress has been made in the understanding of the mathematical structure of DAE's, the analysis of numerical methods applied to DAE systems, the development of robust and efficient mathematical software implementing the numerical methods, and the formulation and solution of DAE systems arising from problems in science and engineering. Many of the results of this research effort are scattered throughout the literature. Our objective in writing this monograph has been to bring together these developments in a unified way, making them more easily accessible to the engineers, scientists, applied mathematicians and numerical analysts who are solving DAE systems or pursuing further research in this area. We have tried not only to present the results on the analysis of numerical methods, but also to show how these results are relevant for the solution of problems from applications and to develop guidelines for problem formulation and effective use of the available mathematical software.

As in every effort of this type, time and space constraints made it impossible for us to address in detail all of the recent research in this field. The research which we have chosen to describe is a reflection of our hope of leaving the reader with an intuitive understanding of those properties of DAE systems and their numerical solution which would in our opinion be most useful in the modeling of problems from science and engineering. In cases where a more extensive description can be found in the literature, we have given references.

Much of our research on this subject has benefitted from collaboration. The DASSL code would never have reached its present state of development without the encouragement of Bob Kee and Bill Winters, whose problems provided us with numerical difficulties that we couldn't possibly have dreamed up ourselves. The direction of our research and the development of DASSL have both benefitted enormously from the many users of DASSL who were generous, or in early versions desperate, enough to share with us their experiences and frustrations. We would like to thank John Betts for introducing us to the difficult nonlinear higher index trajectory problems which inspired much of our research on the solution of higher index systems. Our thinking on the analysis of numerical methods for DAE's has been influenced through our collaborations with Kevin Burrage, Ken Clark, Bill Gear, Björn Engquist, Ben Leimkuhler and Per Lötstedt, with whom we have shared a great many happy hours discussing this subject.

Our friends and colleagues have made many useful suggestions and corrected a large number of errors in early versions of the manuscript. We would like to thank Peter Brown, George Byrne, Alan Hindmarsh, Bob Kee, Michael Knorrenschild, Ben Leimkuhler, Quing Quing Fu and Clement Ulrich for their comments and constructive criticism on various chapters. Claus Führer provided us with such excellent and extensive comments on Section 6.2 that he can virtually be considered a coauthor of that section.

We are grateful to our management at Lawrence Livermore National Laboratory and at The Aerospace Corporation for their support of this project, and for their confidence and patience when it became clear that the effort would take longer to complete than we had originally planned. Very special thanks are due to Don Austin of the Applied Mathematical Sciences Subprogram of the Office of Energy Research, U. S. Department of Energy, and to Lt. Col. James Crowley of the Air Force Office of Scientific Research for their funding and support not only of this project but also of much of the research upon which this monograph is based. Part of this work was performed under the auspices of the U. S. Department of Energy by the Lawrence Livermore National Laboratory under Contract W-7405-Eng-48. Fran Karmitz and Auda Motschenbacher provided much needed expert assistance in typesetting various sections of the manuscript and its revisions.

March 31, 1989 *K. E. Brenan*
S. L. Campbell
L. R. Petzold

Chapter 1

Introduction

1.1 Why DAE's?

Most treatments of ordinary differential equations (ODE's), both analytical and numerical, begin by defining the first order system

$$F(t, y(t), y'(t)) = 0, \qquad (1.1.1)$$

where F and y are vector valued. An assumption that (1.1.1) can be rewritten in the *explicit*, or *normal* form

$$y'(t) = f(t, y(t)) \qquad (1.1.2)$$

is then typically invoked. Thereafter, the theorems and numerical techniques developed concern only (1.1.2). While (1.1.2) will continue to be very important, there has been an increasing interest in working directly with (1.1.1). This monograph will describe the current status of that effort.

Our emphasis is on the numerical solution of (1.1.1) in those cases where working directly with (1.1.1) either has proved, or may prove to be, advantageous. This perspective has affected our choice of techniques and applications. We assume that the reader has some knowledge of traditional numerical methods for ordinary differential equations, although we shall review the appropriate information as needed. We confine ourselves to initial value problems.

If (1.1.1) can, in principle, be rewritten as (1.1.2) with the same state variables y, then it will be referred to as a system of *implicit* ODE's. In this monograph we are especially interested in those problems for which this rewriting is impossible or less desirable. In a system of *differential-algebraic equations*, or DAE's, there are algebraic constraints on the variables. The constraints may appear explicitly as in (1.1.3b) of the system

$$\begin{aligned} F(x', x, y, t) &= 0 & (1.1.3a) \\ G(x, y, t) &= 0, & (1.1.3b) \end{aligned}$$

where the Jacobian of F with respect to x' (denoted by $\partial F/\partial x' = F_{x'}$) is nonsingular, or they may arise because $F_{y'}$ in (1.1.1) is singular. In the latter case, we assume that the Jacobian is always singular. In the problems considered here, difficulties arise from a lower dimensional solution manifold and the dependence of the solution on derivatives of other terms, rather than from the existence of turning or singular points.

There are several reasons to consider (1.1.1) directly, rather than to try to rewrite it as an ODE. First, when physical problems are simulated, the model often takes the form of a DAE depicting a collection of relationships between variables of interest and some of their derivatives. These relationships may even be generated automatically by a modeling or simulation program. The variables usually have a physical significance. Changing the model to (1.1.2) may produce less meaningful variables. In the case of computer-generated or nonlinear models, it may be time consuming or impossible to obtain an explicit model. Parameters are present in many applications. Changing parameter values can alter the relationships between variables and require different explicit models with solution manifolds of different dimensions. If the original DAE can be solved directly, then it becomes easier for the scientist or engineer to explore the effect of modeling changes and parameter variation. It also becomes easier to interface modeling software directly with design software. These advantages enable researchers to focus their attention on the physical problem of interest. There are also numerical reasons for considering DAE's. The change to explicit form, even if possible, can destroy sparsity and prevent the exploitation of system structure. These points will be examined more carefully in later chapters.

The desirability of working directly with DAE's has been recognized for over twenty years by scientists and engineers in several areas. Depending on the area, DAE's have been called singular, implicit, differential-algebraic, descriptor, generalized state space, noncanonic, noncausal, degenerate, semi-state, constrained, reduced order model, and nonstandard systems. In the 1960's and early 1970's there was a study of the analytical theory of linear constant coefficient and some nonlinear systems [174,227]. This work was based on coordinate changes, reductions, and differentiations. The first practical numerical methods for certain classes of DAE's were the backward differentiation formulas (BDF) of [113]. Beginning in the late 1970's, there was a resurgence of interest in DAE's in both the scientific and mathematical literature. New and more robust codes, such as DASSL [200] and LSODI [143] have recently become available. The theoretical understanding of when to expect these codes to work, and what to do when they do not, has improved. There has been increasing computational experience with a wider variety of applications. This monograph deals with these recent developments, many of which have reached a degree of maturity only within the last two or three years.

1.2 Basic Types of DAE's

None of the currently available numerical techniques work for all DAE's. Some additional conditions, either on the structure of the DAE and/or the numerical method, need to be satisfied. One approach to developing a theory for numerical methods is to make technical assumptions that can be hard to verify or understand but enable proofs to be carried out. We prefer, as much as possible, to consider DAE's under various structural assumptions. This approach is more directly related to problem formulation, makes the assumptions easier to verify in many applications, and has proven useful in deriving new algorithms and results. This structural classification will be begun in this section and is continued in more detail in Chapter 2.

Linear constant coefficient DAE's are in the form

$$Ax'(t) + Bx(t) = f(t), \qquad (1.2.1)$$

where A, B are square matrices of real or complex numbers and t is a real variable. We shall usually assume vectors are real for notational convenience but the results are the same for the complex case. The case of rectangular coefficients has been studied [49,111] but we shall not consider it. The general analytical and numerical behavior of (1.2.1) is well understood. However, there is still some research being done on applications of (1.2.1) in the control literature and in the numerical problem of computing the structure of matrix pencils [126,149,242].

While most applications that we have seen have led to either linear constant coefficient or nonlinear DAE's, *linear time varying DAE's*

$$A(t)x'(t) + B(t)x(t) = f(t) \qquad (1.2.2)$$

with $A(t)$ singular for all t, exhibit much of the behavior which distinguishes general DAE's from linear constant coefficient DAE's. At this time, there are techniques and results which seem appropriate for nonlinear systems, but for which complete and rigorous proofs exist only for linear time varying systems. Hence, (1.2.2) represents an important class of DAE's which we will study in order to gain an understanding of general DAE's. We shall also see that it occurs in some applications.

System (1.2.2) is the general, or *fully-implicit* linear time varying DAE. An important special case is the *semi-explicit* linear DAE

$$\begin{aligned} x_1'(t) + B_{11}(t)x_1(t) + B_{12}(t)x_2(t) &= f_1(t) \\ B_{21}(t)x_1(t) + B_{22}(t)x_2(t) &= f_2(t). \end{aligned}$$

The general (or *fully-implicit*) nonlinear DAE

$$F(t, y(t), y'(t)) = 0$$

may be *linear in the derivative*

$$A(t, y(t))y'(t) + f(t, y(t)) = 0. \tag{1.2.3}$$

This system is sometimes referred to as *linearly implicit*. A special case of (1.2.3) is the *semi-explicit nonlinear DAE*

$$\begin{aligned} x_1'(t) &= f_1(x_1(t), x_2(t), t) \\ 0 &= f_2(x_1(t), x_2(t), t). \end{aligned}$$

Depending on the application, we shall sometimes refer to a system as semi-explicit if it is in the form

$$\begin{aligned} F(x'(t), x(t), y(t), t) &= 0 \\ G(x(t), y(t), t) &= 0 \end{aligned}$$

where $F_{x'}$ is nonsingular. Many problems, such as variational problems, lead to semi-explicit systems. We shall see that such systems have properties which may be exploited by some numerical algorithms. However, since it is always possible to transform a fully-implicit linear constant coefficient DAE to a semi-explicit linear constant coefficient DAE by a constant coordinate change, these two cases are not considered separately. While in Chapter 2 we will discuss what type of operations, such as coordinate transformations or premultiplications by nonsingular matrices, may be *safely* applied to a DAE without altering the behavior of a numerical method, it is important to note here that constant coordinate changes are permitted.

1.3 Applications

In this section we will briefly describe several classes of problems where DAE's frequently arise. The categories overlap to some extent, but there are essentially four types of applications that we consider. Our grouping of these applications is based on how the equations are derived rather than on the type of equations that result. Throughout the remainder of this monograph we shall make reference to these problems in order to illustrate the relevance of key concepts. The numerical solution of several specific applications involving DAE's will be developed in more detail in Chapter 6.

1.3.1 Constrained Variational Problems

Variational problems subject to constraints often lead to DAE's. For example, in a constrained mechanical system with position x, velocity u, kinetic energy $T(x, u)$, external force $f(x, u, t)$ and constraint $\phi(x) = 0$, the Euler-Lagrange formulation yields [121]

$$\begin{aligned} x' &= u \\ \frac{d}{dt}\frac{\partial}{\partial u}T(x, u) &= \frac{\partial T}{\partial x} + f(x, u, t) + G^T \lambda \\ 0 &= \phi(x), \end{aligned}$$

1.3. APPLICATIONS

where $G = \partial\phi/\partial x$, and λ is the Lagrange multiplier. This system can be rewritten as

$$\frac{\partial^2 T}{\partial u^2} u' = g(x, u, t) + G^T \lambda \qquad (1.3.1a)$$
$$x' = u \qquad (1.3.1b)$$
$$0 = \phi(x). \qquad (1.3.1c)$$

This DAE system in the unknown variables u, x, and λ is linear in the derivatives. If, as is often the case, $\partial^2 T/\partial u^2$ is a positive definite matrix, then multiplication of (1.3.1a) by $(\partial^2 T/\partial u^2)^{-1}$ converts (1.3.1) to a semi-explicit DAE.

As a classical example of a DAE arising from a variational problem, consider the equations of motion describing a pendulum of length L. If g is the gravitational constant, λ the force in the bar, and (x, y) the cartesian coordinates of the infinitesimal ball of mass one at the end, we obtain the DAE

$$x'' = \lambda x$$
$$y'' = \lambda y - g$$
$$0 = x^2 + y^2 - L^2.$$

Several problems in robotics have been formulated as DAE's utilizing this variational approach [187]. One example is a robot arm moving with an end point in contact with a surface. Using joint coordinates, the motion of this object may be modeled as the DAE

$$M(x)x'' + G(x, x') = u + B^T(x)\lambda$$
$$0 = \phi(x)$$

with $B = \phi_x$, $x \in \mathcal{R}^n$, $\lambda \in \mathcal{R}^m$, $u \in \mathcal{R}^n$. M is the mass matrix, G characterizes the Coriolis, centrifugal and gravitational effects of the robot, u is the input (control) torque vector at the joints, c defines the contact surface, and $B^T\lambda$ is the contact force vector. This system can be converted using the standard substitution $x' = v$ to a DAE which is linear in the derivative. Other variations such as two robots, inertial loads, and moving contact surfaces also fall into this framework [187].

Other examples of DAE's arising from constrained variational problems include optimal control problems with unconstrained controls. In these problems, there is a process given by

$$x' = f(x, u, t) \qquad (1.3.2)$$

and a cost

$$J[x, u] = \int_{t_0}^{t_1} g(x, u, s) ds. \qquad (1.3.3)$$

The problem is to choose the control u in order to minimize the cost (1.3.3) subject to (1.3.2) and certain specified initial or boundary conditions. The variational equations for (1.3.2),(1.3.3) for the fixed time, fixed endpoint problem yield the semi-explicit DAE system

$$\begin{aligned} x' &= f(x,u,t) \\ \lambda' &= -g_x(x,u,t) - f_x^T \lambda \\ 0 &= g_u(x,u,t) + f_u^T \lambda. \end{aligned}$$

One of the most studied special cases of this DAE is the *quadratic regulator problem* with process and cost

$$\begin{aligned} x' &= Ax + Bu \\ J[x,u] &= \int_{t_0}^{t_1} x^T Q x + u^T R u \, ds, \end{aligned} \quad (1.3.4)$$

where A, B, Q, R are matrices with Q, R positive (semi-)definite. In this case, the variational equations become the linear time varying semi-explicit DAE system

$$\begin{aligned} x' &= Ax + Bu \\ \lambda' &= -Qx + A^T \lambda \\ 0 &= Ru + B^T \lambda. \end{aligned} \quad (1.3.5)$$

Often the matrices A, B, Q, and R are constant. Depending on the initial conditions, boundary conditions, and information sought, these DAE's are frequently boundary value problems. Many other variations on these control problems can be found in [12,37].

1.3.2 Network Modeling

In this approach, one starts with a collection of quantities and known, or desired, relationships between them. Electrical circuits are often modeled this way. The circuit is viewed as a collection of devices such as sources, resistors, inductors, capacitors (and more exotic elements such as gyrators, diodes, etc.) lying on branches connected at nodes. The physical quantities of interest are usually taken to be the currents in the branches and the voltages at the nodes (or voltage drops on the branches). These quantities are related by the device laws which are usually in the form of *v-i* characteristics and Kirchoff's node (current) and loop (voltage) laws. These DAE models are often called descriptor or semistate [191] systems in the electrical engineering literature. In an RLC circuit with linear capacitors, resistors, and inductors, the equations will be semi-explicit and often linear constant coefficient. They will be nonlinear if such devices as diodes or nonlinear resistors are included.

1.3. APPLICATIONS

Many of these systems can be written in the form of a DAE which is linear in the derivative [191]

$$Ax' + B(x) = Du$$
$$y = F(x)$$

where A may be singular, u is the vector of inputs, and y is the vector of outputs or observations.

Earlier we discussed constrained variational equations. A somewhat related problem is that of *prescribed path control*. In prescribed path control, one considers the process (or plant) to be given by

$$x' = f(x, u, t) \tag{1.3.6}$$

where x are the state variables and u are the control variables. The goal is to pick u so that the trajectory x follows some prescribed path

$$0 = g(x, u, t). \tag{1.3.7}$$

Frequently u is absent from (1.3.7). As we shall see later, this can impose numerical difficulties in solving the semi-explicit DAE (1.3.6),(1.3.7).

To illustrate the relationship between this problem and our earlier constrained optimization problem, consider a robotic control problem. If a contact point is moving along a surface, then the constraint is imposed by the surface. The surface exerts forces on the robot and the problem would most likely be modeled by the variational approach. If, however, the robot were moving freely through the workspace and the prescribed path (1.3.7) was being specified to avoid collision with fixed objects, then the problem would most likely be modeled by taking the free dynamics (1.3.6) and imposing the constraint (1.3.7). In Chapter 2 we shall define a fundamental concept called the index of a DAE. Generally, the higher the index, the more difficult the problem is numerically. The prescribed path control of robot arms with flexible joints [88,90] leads to the highest index problems we have seen to date.

In *trajectory* prescribed path control problems, the DAE system models a vehicle flying in space when algebraic path constraints are imposed on its trajectory. A particular example of this type of DAE concerned the design of a safe reentry profile for the space shuttle [32]. In order to insure the shuttle is able to survive the reentry, mission restrictions concerning heating constraints and structural integrity requirements must be evaluated with respect to the forces acting on the vehicle. A particular trajectory profile is selected inside an acceptable corridor of trajectories which has been determined by analyzing numerous physical constraints such as the vehicle's maximum load factors, maximum surface temperature constraints at a set of control points, and equilibrium glide boundaries. The equations modeling the shuttle flying along this reentry profile describe a DAE system in the form (1.3.6),(1.3.7). The differential equations (1.3.6) include the vehicle's translational equations of motion.

The trajectory profile is described by one additional algebraic equation of the form
$$g(x,t) = 0, \tag{1.3.8}$$
where the state variables x describe the vehicle's position and velocity. The bank angle β is selected as the control variable u. The semi-explicit DAE (1.3.6),(1.3.8) exhibits several features that will be discussed in Chapters 2 and 6. Note that because of the many nonlinearities involved in this model problem, it is not easy to convert this DAE into an explicit ODE, even locally.

Another type of prescribed path control problem arises when there are invariants (i.e., relations which hold along any solution trajectory) present in the solution to a system of ODE's. For example, these invariants may be equalities (or inequalities in general) describing the conservation of total energy, mass, or momentum of a system. Maintaining solution invariants in the numerical solution of ODE's can pose difficulties for numerical ODE methods [223], but may be essential to the stability of the system. Problems of this type may be formulated as systems of DAE's where the invariants are imposed as algebraic constraints [117].

There is a large engineering literature devoted to the properties of the system
$$\begin{aligned} x' &= Ax + Bu \\ y &= Cx + Du \end{aligned}$$
where A, B, C, D are constant matrices, x is the state, u is the control or input, and y is the output or vector of observations. Notice that if we consider a desired y as known and want to find x or u, then this system can be viewed as the linear constant coefficient DAE in x and u
$$\begin{bmatrix} I & 0 \\ 0 & 0 \end{bmatrix} \begin{bmatrix} x \\ u \end{bmatrix}' = \begin{bmatrix} A & B \\ C & D \end{bmatrix} \begin{bmatrix} x \\ u \end{bmatrix} + \begin{bmatrix} 0 \\ -y \end{bmatrix}.$$

Our final network example is derived from the equations describing a chemical reactor [199]. A first-order isomerisation reaction takes place and the heat generated is removed from the system through an external cooling circuit. The relevant equations are

$$\begin{aligned} C' &= K_1(C_0 - C) - R & \text{(1.3.9a)} \\ T' &= K_1(T_0 - T) + K_2 R - K_3(T - T_C) & \text{(1.3.9b)} \\ 0 &= R - K_3 \exp\left(\frac{-K_4}{T}\right) C & \text{(1.3.9c)} \\ 0 &= C - u. & \text{(1.3.9d)} \end{aligned}$$

Here C_0 and T_0 are the (assumed) known feed reactant concentration and feed temperature, respectively. C and T are the corresponding quantities in the product. R is the reaction rate per unit volume, T_C is the temperature of the

1.3. APPLICATIONS

cooling medium (which can be varied), and the K_i are constants. We shall consider two variations of this problem later. The simpler one is (1.3.9a)-(1.3.9c) with T_C known and the state variables C, T, R. The more interesting one is (1.3.9a)-(1.3.9d) where (1.3.9d) is a specified desired product concentration and we want to determine the T_C (control) that will produce this C. In this case, we obtain a semi-explicit DAE system with state variables C, T, R, T_C. Even though these two problems turn out to be different in several respects (for example, in the dimension of their solution manifolds), they can be studied from the same DAE model. This example again illustrates the flexibility of working with a problem which is formulated as a DAE.

1.3.3 Model Reduction and Singular Perturbations

In a given model, there may be various small parameters. In an attempt to simplify the model or obtain a first order approximation to its behavior, these parameters may be set equal to zero. The resulting system is often a DAE.

In the classical singular perturbation problem (1.3.10) with $0 < \epsilon \ll 1$,

$$\begin{aligned} x' &= f(x,y,t,\epsilon) \\ \epsilon y' &= g(x,y,t,\epsilon) \end{aligned} \qquad (1.3.10)$$

setting $\epsilon = 0$ leads to the reduced order model

$$\begin{aligned} x' &= f(x,y,t,0) \\ 0 &= g(x,y,t,0). \end{aligned} \qquad (1.3.11)$$

This semi-explicit DAE may be used if parasitics are to be ignored. Even if the solution of (1.3.10) is sought for all $t \geq 0$, the DAE (1.3.11) can often be solved and the solution added to a boundary layer correction term corresponding to a fast time scale to obtain a solution.

In general, there may be several small parameters, and the original equations may also be a DAE. As an example, consider the circuit (known as a loaded degree-one Hazony section under small loading) described in [51,102]. This circuit has time-invariant linear resistors, capacitors, a current source, and a gyrator. A gyrator is a 2-port device for which the voltage and current at one port are multiples of the current and voltage at the other port. The resistances R_i are large since they model parasitic terms, and similarly the capacitances are small. Letting $G_i = 1/R_i$, $G = 1/R$, g the gyration conductance, $u = i_S$ the current source, $y = v_2$, $x = [v_1, v_2, i_3]^T$, we obtain, after some algebraic simplification, the linear constant coefficient DAE

$$\begin{bmatrix} C_1 & 0 & 0 \\ 0 & C_2 & 0 \\ C & -C_2 & 0 \end{bmatrix} x' + \begin{bmatrix} G_1 & g & 1 \\ g & G+G_2 & -1 \\ G_3 & -G_3 & -1 \end{bmatrix} x = \begin{bmatrix} 1 \\ 0 \\ 0 \end{bmatrix} u \qquad (1.3.12a)$$

$$y = \begin{bmatrix} 0 & 1 & 0 \end{bmatrix} x. \qquad (1.3.12b)$$

Here we assume that $G \geq 0$, $g > 0$, $C > 0$. In this problem there are five small parasitic parameters G_1, G_2, G_3, C_2, C_1 and one small model parameter G. The state dimension changes with G but all variations are easily examined within the one DAE model (1.3.12) [51].

Another example of DAE's arising from singular perturbation problems appears in cheap control problems. A simple example is the quadratic regulator problem mentioned earlier, where the control weighting matrix R depends on a small parameter $R(\epsilon)$ and $R(0)$ is either singular or of lower rank.

Problems such as (1.3.10) are often referred to as stiff differential equations. It is well known that the solution of stiff differential equations, even for explicit ODE's, requires a special choice of numerical method. For example, explicit methods require exceedingly small stepsizes for stability reasons, so implicit methods are generally employed. The $\epsilon = 0$ problem results when the stiffness is pushed to the limit. It is not surprising then that parallels will be found in the following chapters between the theory for the numerical solution of DAE's and that for stiff differential equations.

1.3.4 Discretization of PDE's

Solving partial differential equations (PDE's) can lead to DAE's. We shall consider the method of lines and moving grids.

Numerical methods for solving PDE's usually involve the replacement of all derivatives by discrete difference approximations. The *method of lines* (MOL) does this also, but in a special way that takes advantage of existing software. For parabolic PDE's, the typical MOL approach is to discretize the spatial derivatives, for example by finite differences, and thus convert the PDE system into an ODE initial value problem.

There are two important advantages to the MOL approach. First, it is computationally efficient. The ODE software takes on the burden of time discretization and of choosing the time steps in a way that maintains accuracy and stability in the evolving solution. Most production ODE software is written to be robust and computationally efficient. Also, the person using a MOL approach has only to be concerned with discretizing spatial derivatives, thus reducing the work required to write a computer program.

Many MOL problems lead, either initially or with slight modification, to an explicit ODE. However, many well posed problems of practical interest are more easily handled as a DAE. As a simple first example of the MOL consider the heat equation

$$\frac{\partial y}{\partial t} = \frac{\partial^2 y}{\partial x^2}$$

defined on the region $t \geq 0$, $0 \leq x \leq 1$ with boundary conditions given for $y(t,0)$ and $y(t,1)$, and initial conditions given for $y(0,x)$. Taking a uniform spatial mesh of Δx, and mesh points $x_j = (j+1)\Delta x$, $1 \leq j \leq (1/\Delta x) - 1 = N$, and using centered differences, we obtain the semi-explicit DAE in the variables

1.3. APPLICATIONS

$y_i(t) = y(t, x_i)$

$$y'_j - \frac{y_{j-1} - 2y_j + y_{j+1}}{(\Delta x)^2} = 0, \quad j = 2, \ldots, N-1$$

$$y_1 - y(t, 0) = 0$$

$$y_N - y(t, 1) = 0.$$

This particular problem is easily reformulated as an ODE, but this is not always the case.

As a more interesting example, consider the following equations for ignition of a single component nonreacting gas in a closed cylindrical vessel in Lagrangian coordinates [151]

$$\frac{\partial T}{\partial t} - \frac{1}{\rho c_p} \frac{\partial p}{\partial t} = \frac{1}{c_p} \frac{\partial}{\partial \psi} \left(\rho r^2 \lambda \frac{\partial T}{\partial \psi} \right) \tag{1.3.13a}$$

$$0 = \frac{\partial r}{\partial \psi} - \frac{1}{\rho r} \tag{1.3.13b}$$

$$0 = \frac{\partial p}{\partial \psi} \tag{1.3.13c}$$

$$0 = p - \rho \frac{RT}{W} \tag{1.3.13d}$$

with conservation of mass

$$\int_0^R \rho r \, dr = \text{constant}.$$

Since (1.3.13d) can be used to evaluate ρ wherever it occurs, we consider the DAE obtained by discretizing (1.3.13a)-(1.3.13c) with state variables r = spatial coordinate, T = temperature, and p = pressure. The boundary conditions are at the center ($\psi = 0$)

$$r = 0, \quad \frac{\partial T}{\partial \psi} = 0$$

and at the vessel boundary ($\psi = \psi_R$)

$$r = R, \quad T = T_w.$$

Notice that upon discretization, the only time derivatives appear in the equations derived from (1.3.13a) and none appear from (1.3.13b),(1.3.13c). The latter become algebraic equations. Also there are only two boundary conditions on r and T, but none on p. Thus not only is the discretized system naturally a DAE, but the boundary conditions on p are implicit. As will be discussed in Chapter 6, this situation is easily handled by existing DAE codes.

There is another way that utilizing DAE's is helpful with MOL. The analytic solution of a system of PDE's may satisfy an important physical property, such as incompressibility. However, the numerical solution may not. One way to handle this in the MOL approach is to add the physical property as an algebraic condition on the ODE (or DAE) in the spatial discretization. The numerical solution of this new DAE will then still possess the desired physical property.

Our final fixed-mesh MOL example is the flow of an incompressible viscous fluid described by the Navier-Stokes equations

$$\frac{\partial u}{\partial t} + (u \cdot \nabla)u = -\nabla p + \gamma \nabla^2 u \qquad (1.3.14a)$$
$$\nabla \cdot u = 0, \qquad (1.3.14b)$$

where u is the velocity in two or three space dimensions, p is the scalar pressure, and γ is the kinematic viscosity. Equation (1.3.14a) is the momentum equation and (1.3.14b) is the incompressibility condition. After spatial discretization of (1.3.14) with a finite difference or finite element method, the vectors $U(t)$, $P(t)$ approximating $u(t,x)$, $p(t,x)$ in the domain of interest satisfy [128]

$$MU' + (K + N(U))U + CP = f(U, P)$$
$$C^T U = 0.$$

In this DAE the mass matrix M is either the identity if finite differences are used (and the resulting DAE is semi-explicit) or symmetric positive definite in the case of finite elements. In the latter case the DAE could be made semi-explicit by multiplying by M^{-1} but the sparsity of the coefficients of the DAE would be lost. The discretization of the operator ∇ is C, and the forcing function f comes from the boundary conditions.

In many applications, such as combustion modeling, the use of a fixed spatial mesh leads either to poor accuracy or to a fine mesh and large computational effort. One way to circumvent this difficulty is to let the mesh points change with the time t. Thus at time t, the discretized system has two sets of variables, the mesh points $\{x_j(t)\}$ and the values of u at these meshpoints $\{u(x_j,t)\}$. Suppose that we initially have the PDE

$$F(u_t, u, u_x, u_{xx}) = 0, \qquad (1.3.15)$$

and the grid velocity x' is chosen to minimize the time rate of change of u and x in the new coordinates. That is,

$$\min_{x'}(\|u'\|^2 + \alpha \|x'\|^2), \qquad (1.3.16)$$

where α is a positive scaling parameter. Then the DAE that describes this adaptive moving mesh MOL is the spatial discretization of

$$F(u' + u_x x', u, u_x, u_{xx}) = 0$$
$$\alpha x' + u' u_x = 0. \qquad (1.3.17)$$

1.4. OVERVIEW 13

Other criteria can be used to determine x' in a moving mesh formulation. However, (1.3.16) is the one discussed in more detail in Chapter 6. Note that even if (1.3.15) were explicit

$$u_t = f(u, u_x, u_{xx})$$

then (1.3.17) would only be linear in the derivative and not semi-explicit.

1.4 Overview

This chapter has introduced the idea of a DAE, suggested some of the reasons why DAE's are important, and given some examples of applications which are naturally formulated as DAE's.

DAE's differ in several key aspects from explicit ODE's. However, problems from applications possess different kinds of structure which can be exploited for the numerical solution. Chapter 2 lays the theoretical foundation necessary for understanding the material in the remaining chapters. The analytical properties of the solutions of DAE's are developed. The equation structure to be exploited in the numerical methods is defined and discussed.

The first numerical methods to be successfully used on DAE's were linear multistep methods such as backward differentiation formulas (BDF) [113]. Chapter 3 develops the theory for these methods. In particular, stability and convergence properties are examined. General comments on the applicability and limitations of these methods are given.

There has recently developed a better understanding of the use of one-step methods such as implicit Runge-Kutta and extrapolation in the numerical solution of DAE's. These methods are examined in Chapter 4.

Chapter 5 discusses the available general purpose codes for solving DAE's. Considerations in developing future software are addressed.

Chapter 6 examines the numerical solution of DAE's from applications to illustrate all of the preceding. The important role of problem formulation is addressed. Guidelines for the numerical solution of these problems are developed with respect to computational experiences with specific examples.

Chapter 2

Theory of DAE's

2.1 Introduction

DAE's differ in several key aspects from ODE's. In this chapter we develop the theory needed to understand these differences and their implications for DAE models. As we shall see in Chapters 3 and 4, classical numerical methods such as backward differentiation formulas and implicit Runge-Kutta methods cannot be used to approximate the solutions to all DAE's. Thus it is important to be able to recognize those classes of problems for which codes based on these methods will work. The most useful approach to date has been the development of various (structural) forms of DAE's. These forms have the advantages that they arise in many applications and are often relatively easy to identify. Our focus in this monograph is on the numerical solution of DAE's, but the various forms have also proved useful in developing general analytic properties of DAE's.

Historically, the understanding of DAE's has progressed from simpler cases to more general ones. In particular, the definitions of key concepts such as index and solvability have evolved over several years. In Section 2.2 we give the definitions of these concepts which have proved to be the most useful in the general nonlinear setting. In order to better understand the difference between DAE's and ODE's, we then consider a sequence of increasingly general systems starting with linear constant coefficient DAE's in Section 2.3. The definitions may appear somewhat different from those used in the original development of these special cases [49,50], but the connections will be pointed out as we go along. Next, linear time varying DAE's are considered in Section 2.4, while nonlinear systems of DAE's are discussed in Section 2.5. Many interesting analytical and algebraic results not directly related to the numerical solution of DAE's have been omitted. However, citations of the relevant literature are provided for the interested reader.

2.2 Solvability and the Index

A function $y(t)$ is a (classical) solution of the general nonlinear DAE

$$F(t, y, y') = 0 \tag{2.2.1}$$

on an interval $\mathcal{I}$, if y is continuously differentiable on $\mathcal{I}$ and satisfies (2.2.1) for all $t \in \mathcal{I}$. As we shall see shortly, being the solution of a DAE may actually imply that some portions of y are more differentiable than other portions which may be only continuous. One development of the theory allowing the latter for some DAE's is given by Griepentrog and März [132]. To avoid becoming too technical at this point, we merely assume that y is smooth enough to satisfy (2.2.1). DAE's can exhibit all the behavior of ODE's plus additional behavior such as bifurcation of solutions. In this monograph, we are concerned with the case where the solutions exist and are uniquely defined on the interval of interest; however, not all initial values for y admit a smooth solution. This concept of *solvablility* is made more precise in the next definition.

Definition 2.2.1 *Let $\mathcal{I}$ be an open subinterval of $\mathcal{R}$, Ω a connected open subset of $\mathcal{R}^{2m+1}$, and F a differentiable function from Ω to $\mathcal{R}^m$. Then the DAE (2.2.1) is solvable on $\mathcal{I}$ in Ω if there is an r-dimensional family of solutions $\phi(t,c)$ defined on a connected open set $\mathcal{I} \times \tilde{\Omega}$, $\tilde{\Omega} \subset \mathcal{R}^r$, such that:*

1. *$\phi(t,c)$ is defined on all of $\mathcal{I}$ for each $c \in \tilde{\Omega}$*
2. *$(t, \phi(t,c), \phi'(t,c)) \in \Omega$ for $(t,c) \in \mathcal{I} \times \tilde{\Omega}$*
3. *If $\psi(t)$ is any other solution with $(t, \psi(t), \psi'(t)) \in \Omega$, then $\psi(t) = \phi(t,c)$ for some $c \in \tilde{\Omega}$*
4. *The graph of ϕ as a function of (t,c) is an $r+1$-dimensional manifold.*

The definition says that locally there is an r-dimensional family of solutions. At any time $t_0 \in \mathcal{I}$, the initial conditions form an r-dimensional manifold $\phi(t_0, c)$ and r is independent of t_0. The solutions are a continuous function of the initial conditions on this manifold (or equivalently of c). Since the solutions exist for all $t \in \mathcal{I}$ and (3) holds, there are no bifurcations of these solutions. For an implicit ODE such as $\tan y' = y$, $y(0) = 0$, there are countably many solutions given by the solutions of $y' = \tan^{-1} y$, $y(0) = 0$, for different branches of $\tan^{-1}$. In this simple example, $\tilde{\Omega}$ would correspond to choosing a particular branch of $\tan^{-1}$, say by taking $\Omega = \mathcal{R} \times (-\frac{\pi}{2}, \frac{\pi}{2}) \times \mathcal{R}$.

A property known as the *index* plays a key role in the classification and behavior of DAE's. In order to motivate the definition that follows, consider the special case of a semi-explicit DAE

$$\begin{align}
x' &= f(x, y, t) \tag{2.2.2a} \\
0 &= g(x, y, t). \tag{2.2.2b}
\end{align}$$

2.2. SOLVABILITY AND THE INDEX

If we differentiate the constraint equation (2.2.2b) with respect to t, we get

$$x' = f(x,y,t) \tag{2.2.3a}$$
$$g_x(x,y,t)x' + g_y(x,y,t)y' = -g_t(x,y,t). \tag{2.2.3b}$$

If g_y is nonsingular, the system (2.2.3) is an implicit ODE and we say that (2.2.2) has index one. If this is not the case, suppose that with algebraic manipulation and coordinate changes we can rewrite (2.2.3) in the form of (2.2.2) but with different x, y. Again we differentiate the constraint equation. If an implicit ODE results, we say that the original problem has index two. If the new system is not an implicit ODE, we repeat the process. The number of differentiation steps required in this procedure is the index.

Definition 2.2.2 *The minimum number of times that all or part of* (2.2.1) *must be differentiated with respect to t in order to determine* y' *as a continuous function of* y, t, *is the* index *of the DAE* (2.2.1).

It should be stressed that we are not recommending the previous series of differentiations and coordinate changes as a general solution procedure for DAE's, even though some such manipulation is frequently helpful. Rather the number of such differentiation steps that would be required in theory turns out to be an important quantity in understanding the behavior of numerical methods. In optimal control theory [12], there is a similar concept known as the *order* of a singular arc which is related to the number of differentiations needed to explicitly get the control to appear in the algebraic part of the variational equations. The definition of the index is actually much more general than our discussion suggests, since it makes sense for general solvable DAE's which can not even be put in semi-explicit form. Also, it is not obvious that the sequence of differentiations and coordinate changes given in the motivational discussion requires the same number of steps as just differentiating the original DAE several times and then solving for y'. These latter points will be discussed in detail in Section 2.5.

This definition suggests several other points that will be developed in the sections to follow, but are important enough to comment on now. First, if (2.2.1) has index ν, then its solutions also satisfy an ODE

$$y' = G(t,y), \tag{2.2.4}$$

where G is a function of partial derivatives of F. Of course, not all solutions of (2.2.4) are solutions of the original DAE. However, (2.2.4) is useful both theoretically and in some general numerical procedures. This ODE also suggests that in general the solution of a DAE can involve derivatives of both input functions and system coefficients. This characteristic will turn out to have several ramifications for numerical procedures.

What we have defined as the index was originally called the *global index* [122]. Note also that an implicit ODE has *index zero* according to this definition. We will see that DAE's with index zero and one are generally much simpler to understand than DAE's with index two or higher. Often, we will refer to DAE's with index greater than one simply as *higher index* systems.

2.3 Linear Constant Coefficient DAE's

Linear constant coefficient DAE's are the best understood class of DAE systems. They have been studied for many years, particularly in the electrical engineering and control literature [49,163]. We shall develop just that portion of the theory needed to understand the numerical methods to be discussed later and to begin to appreciate how DAE's differ from ODE's.

Consider the *linear constant coefficient DAE*

$$Ax' + Bx = f \qquad (2.3.1)$$

where A, B are $m \times m$ matrices. If λ is a complex parameter, then $\lambda A + B$ is called a *matrix pencil*. If we let $x = Qy$, and premultiply by P in (2.3.1), where P, Q are $m \times m$ nonsingular matrices, (2.3.1) becomes

$$PAQy' + PBQy = Pf \qquad (2.3.2)$$

and the new pencil is $\lambda PAQ + PBQ$. The nature of the solutions to (2.3.1) is thus determined by the canonical form of the pencil $\lambda A + B$ under the transformations P, Q. The most detailed description is given by the *Kronecker canonical form* [111]. However, this form is rarely computed in practice and we shall develop a weaker version that suffices for our purposes. Since our primary interest is the numerical solution of nonlinear DAE's, we shall not discuss the actual computation of solutions of linear constant coefficient DAE's. The interested reader is referred to [126,149,242].

If the determinant of $\lambda A + B$, denoted $\det(\lambda A + B)$, is not identically zero as a function of λ, then the pencil is said to be *regular*. Solvability as defined in Section 2.2 can be difficult to determine for general DAE's, but for (2.3.1) there is a nice characterization.

Theorem 2.3.1 *The linear constant coefficient DAE (2.3.1) is solvable if and only if $\lambda A + B$ is a regular pencil.*

For the remainder of this section we shall assume that $\lambda A + B$ is a regular pencil. Recall that a matrix N is said to have nilpotency k if $N^k = 0$ and $N^{k-1} \neq 0$. The key structure theorem for (2.3.1) which follows from the Kronecker form, or which may be derived directly [49] is:

Theorem 2.3.2 *Suppose that $\lambda A + B$ is a regular pencil. Then there exist nonsingular matrices P, Q such that*

$$PAQ = \begin{bmatrix} I & 0 \\ 0 & N \end{bmatrix}, \quad PBQ = \begin{bmatrix} C & 0 \\ 0 & I \end{bmatrix} \qquad (2.3.3)$$

2.3. LINEAR CONSTANT COEFFICIENT DAE'S

where N is a matrix of nilpotency k and I is an identity matrix. If $N = 0$, then define $k = 1$. In the special case that A is nonsingular, we take $PAQ = I$, $PBQ = C$ and define $k = 0$. If $\det(\lambda A + B)$ is identically constant, then (2.3.3) simplifies to $PAQ = N$, $PBQ = I$.

The degree of nilpotency of N in this theorem, namely the integer k, is the *index of the pencil* $\lambda A + B$. It is also the index of the DAE as described in Section 2.2 (i.e., $k = \nu$). Alternately, the *index of a matrix* M can be defined as the least nonnegative integer $\hat{\nu}$ such that $\mathcal{N}(M^{\hat{\nu}}) = \mathcal{N}(M^{\hat{\nu}+1})$ where $\mathcal{N}$ denotes the null space [49]. In the case when $\det(\lambda A + B)$ is constant, we call the pencil a *nilpotent pencil*. Before pursuing these relationships any further, let us examine the solutions of (2.3.1).

Suppose that the coordinate changes P, Q provided by this theorem are applied to the DAE (2.3.1) to obtain (2.3.2), which by (2.3.3) is

$$y_1' + C y_1 = f_1 \qquad (2.3.4a)$$
$$N y_2' + y_2 = f_2. \qquad (2.3.4b)$$

The first equation (2.3.4a) is an ODE and a solution exists for any initial value of y_1 and any continuous forcing function f_1. Equation (2.3.4b) has only one solution (initial values for y_2 are completely determined). The quickest way to see this is to write (2.3.4b) as

$$(ND + I)y_2 = f_2, \qquad (2.3.5)$$

where $D = d/dt$. But $(ND)^k = 0$ and (2.3.5) imply that

$$y_2 = (ND + I)^{-1} f_2 = \sum_{i=0}^{k-1} (-1)^i N^i f_2^{(i)} \qquad (2.3.6)$$

where $f_2^{(i)} = d^i f_2 / dt^i$. This calculation illustrates several key points about DAE's:

1. The solution of (2.3.1) can involve derivatives of order $k-1$ of the forcing function f if the DAE is higher index.

2. Not all initial conditions of (2.3.1) admit a smooth solution if $k \geq 1$. Those that do admit solutions are called *consistent initial conditions*.

3. Higher index DAE's can have hidden algebraic constraints.

This last point is important but not as immediately obvious as the first two. Consider the following semi-explicit DAE

$$x_1' + x_3 = f_1 \qquad (2.3.7a)$$
$$x_2' + x_1 = f_2 \qquad (2.3.7b)$$
$$x_2 = f_3. \qquad (2.3.7c)$$

There is one explicit algebraic constraint which is equation (2.3.7c). However, this DAE has only the one solution

$$\begin{aligned} x_1 &= f_2 - f_3' \\ x_2 &= f_3 \\ x_3 &= f_1 - f_2' + f_3'' \end{aligned}$$

because there are two additional implicit algebraic constraints in the original DAE.

In Section 2.2 we defined the index of a DAE as the number of differentiations needed to determine x'. We will now illustrate this definition and relate it to Theorem 2.3.2. Assume that (2.3.1) is solvable with A singular but nonzero. Then there exists a nonsingular P so that premultiplication of (2.3.1) by P gives

$$\begin{bmatrix} A_1 & A_2 \\ 0 & 0 \end{bmatrix} x' + \begin{bmatrix} B_1 & B_2 \\ B_3 & B_4 \end{bmatrix} x = f \qquad (2.3.8)$$

where $\begin{bmatrix} A_1 & A_2 \end{bmatrix}$ has full row rank. The explicit algebraic constraints are $B_3 x_1 + B_4 x_2 = f_2$. If we differentiate this constraint, then (2.3.8) becomes

$$\begin{bmatrix} A_1 & A_2 \\ B_3 & B_4 \end{bmatrix} x' + \begin{bmatrix} B_1 & B_2 \\ 0 & 0 \end{bmatrix} x = \begin{bmatrix} f_1 \\ f_2' \end{bmatrix}. \qquad (2.3.9)$$

If the new coefficient of x' in (2.3.9) is singular, then this procedure is repeated. Using the Kronecker form, it is straightforward to show

Theorem 2.3.3 *Suppose that $Ax' + Bx = f$ is a constant coefficient solvable DAE with index ≥ 1. Then each iteration of solving for the explicit constraints and differentiating these constraints reduces the index by one so that after ν differentiations, the DAE has been reduced to an ODE. Furthermore, the index of the DAE is ν, which is also the degree of nilpotency of N (equivalently, the index of N) in (2.3.3).*

It is important to keep in mind that only some of the solutions of the lower index problem derived by differentiation are solutions of the original DAE. To illustrate, if this procedure is applied to (2.3.7) we get the ODE

$$\begin{aligned} x_1' &= f_2' - f_3'' \\ x_2' &= f_3' \\ x_3' &= f_1' - f_2'' + f_3''' \end{aligned}$$

which has many additional solutions (namely, solutions for arbitrary initial values providing the forcing functions are sufficiently smooth).

When we study numerical methods, we will see that the matrix $(A + \lambda B)^{-1}$ plays a key role. Furthermore, the conditioning of this matrix is closely related to the system's index.

2.3 LINEAR CONSTANT COEFFICIENT DAE'S

Theorem 2.3.4 *Suppose that* $Ax' + Bx = f$ *is a solvable linear constant coefficient DAE with index* $\nu \geq 1$. *Then* $(A + \lambda B)^{-1}$ *has a pole of order* ν *at* $\lambda = 0$ *and* $(A + \lambda B)^{-1}A$ *has a pole of order* $\nu - 1$ *at* $\lambda = 0$.

Proof. Let P, Q be nonsingular matrices so that (2.3.3) holds. Then

$$\begin{aligned}
(A + \lambda B)^{-1} &= Q(PAQ + \lambda PBQ)^{-1}P \\
&= Q \begin{bmatrix} (I + \lambda C)^{-1} & 0 \\ 0 & (N + \lambda I)^{-1} \end{bmatrix} P \\
&= Q \begin{bmatrix} (I + \lambda C)^{-1} & 0 \\ 0 & \frac{1}{\lambda}\sum_{i=0}^{\nu-1} \frac{(-N)^i}{\lambda^i} \end{bmatrix} P.
\end{aligned}$$

Also,

$$\begin{aligned}
(A + \lambda B)^{-1}A &= Q(PAQ + \lambda PBQ)^{-1}PAQQ^{-1} \\
&= Q \begin{bmatrix} (I + \lambda C)^{-1} & 0 \\ 0 & \frac{1}{\lambda}\sum_{i=0}^{\nu-1} \frac{(-N)^i}{\lambda^i} \end{bmatrix} \begin{bmatrix} I & 0 \\ 0 & N \end{bmatrix} Q^{-1} \\
&= Q \begin{bmatrix} (I + \lambda C)^{-1} & 0 \\ 0 & \frac{1}{\lambda}\sum_{i=0}^{\nu-2} \frac{(-N)^{i+1}}{\lambda^i} \end{bmatrix} Q^{-1}. \quad \square
\end{aligned}$$

There are several other aspects of linear constant coefficient DAE's that are important, but they are also true for the linear time varying case and will be discussed in the next section.

We conclude this section with a few comments on nilpotent matrices that will be useful in what follows. A square matrix is *strictly upper (lower) triangular* if all the entries on and below (above) the main diagonal are zero. The next theorem summarizes several key facts about nilpotent matrices.

Theorem 2.3.5 *If N is a nilpotent matrix, then there exist nonsingular matrices P_1 and P_2 such that $P_1 N P_1^{-1}$ is strictly lower triangular and $P_2 N P_2^{-1}$ is strictly upper triangular. If N has index ν, then P_1 and P_2 can be chosen so that $P_1 N P_1^{-1}$ and $P_2 N P_2^{-1}$ are $\nu \times \nu$ block matrices which are strictly lower and upper block triangular, respectively. Conversely, any strictly upper, or lower, triangular square matrix is nilpotent.*

A property of a matrix is a *structural property* if all the nonzero entries of the matrix can be replaced by independent variables and the property still holds on an open dense set of values for these variables. For the property of nilpotency we have

Theorem 2.3.6 *A square matrix N is structurally nilpotent if and only if there is a permutation matrix P so that PNP^{-1} is strictly triangular.*

Structural characteristics will turn out to be important both in the study of analytical properties of DAE's and in the convergence and stability properties of numerical methods. Often, the structure of the DAE systems arising in practical problems naturally includes 'hard' zeroes (e.g., semi-explicit DAE's). The index of many of these systems can be determined by examining this zero structure. In fact, Duff and Gear [101] give an algorithm based only on the system structure that determines if the structural index is one, two, or greater. Later we will also see how the convergence and stability properties of numerical methods are dependent on the structure of the system.

2.4 Linear Time Varying DAE's

2.4.1 Solvability and the Index

The *linear time varying DAE*

$$A(t)x'(t) + B(t)x(t) = f(t) \tag{2.4.1}$$

defined on the interval $\mathcal{I}$ is important in understanding general DAE's. It exhibits most of the behavior found in the nonlinear case that is not already present in the constant coefficient case, yet the linearity facilitates the analysis. In spite of this, many aspects of the theory and numerical solution of (2.4.1) have only been resolved within the last few years. Some questions remain open.

The basis of our understanding of the linear constant coefficient DAE in the last section was the form (2.3.3) for a regular pencil. Similarly, forms will play a crucial role in our study of (2.4.1). The ensuing discussion will develop several forms for DAE's. Most of these forms have natural analogues for nonlinear systems that will be discussed in the next section. These forms have two roles. Some of them tend to occur in applications and give the problem structure that can be exploited numerically. Other forms are convenient for proving various numerical and analytical properties of DAE's and for increasing our understanding of their behavior.

Recall from Section 2.2 the general definitions of solvability and the index, here phrased in terms of (2.4.1).

Definition 2.4.1 *The system* (2.4.1) *with A, B $m \times m$ matrices is* solvable *on the interval $\mathcal{I}$ if for every m-times differentiable f, there is at least one continuously differentiable solution to* (2.4.1). *In addition, solutions are defined on all of $\mathcal{I}$ and are uniquely determined by their value at any $t \in \mathcal{I}$.*

Definition 2.4.2 *Suppose that* (2.4.1) *is solvable on $\mathcal{I}$. The* index *of* (2.4.1) *is the minimum number of differentiations (for all f) needed to uniquely determine x' as a continuous function of x, t.*

2.4. LINEAR TIME VARYING DAE'S

A rigorous discussion of the index is somewhat technical in nature and will be done at the end of this section after we have developed a better understanding of linear time varying DAE's.

To motivate the rest of this section, suppose that we wish to solve (2.4.1) by the implicit Euler method starting at time t_0 with constant stepsize h. Let $t_n = t_0 + nh$, x_n be the estimate for $x(t_n)$, and $c_n = c(t_n)$ for $c = f, A, B$. Then the implicit Euler method applied to (2.4.1) gives

$$A_n \frac{x_n - x_{n-1}}{h} + B_n x_n = f_n$$

or

$$(A_n + hB_n)x_n = A_n x_{n-1} + hf_n. \qquad (2.4.2)$$

In order for (2.4.2) to uniquely determine x_n given x_{n-1}, we need $A(t_n) + hB(t_n)$ to be nonsingular for small h. Thus we need regularity of the pencil $\lambda A(t) + B(t)$ for each $t \in \mathcal{I}$. In this case we say that (2.4.1) is a *regular DAE* on $\mathcal{I}$.

Solvability and regularity were equivalent for linear constant coefficient DAE's. This is no longer the case for linear time varying DAE's. In fact, the concepts turn out to be independent.

Example 2.4.1 Let

$$A(t) = \begin{bmatrix} 1 & t \\ 0 & 0 \end{bmatrix}, \quad B(t) = \begin{bmatrix} 0 & 0 \\ 1 & t \end{bmatrix}.$$

Then $\lambda A(t) + B(t)$ is singular for all λ, t so that $Ax' + Bx = f$ is not a regular DAE. Let $x = \begin{bmatrix} t & 1 \\ -1 & 0 \end{bmatrix} y$. Then $Ax' + Bx = f$ becomes

$$\begin{bmatrix} 0 & 1 \\ 0 & 0 \end{bmatrix} y' + \begin{bmatrix} 1 & 0 \\ 0 & 1 \end{bmatrix} y = f,$$

which is a solvable linear constant coefficient index two DAE and, in particular, $Ax' + Bx = f$ is solvable.

This example also shows that the numerical method (2.4.2) cannot work for all solvable systems even if we restrict ourselves to index two or less. The next example shows that regularity does not imply solvability.

Example 2.4.2 Let

$$A = \begin{bmatrix} -t & t^2 \\ -1 & t \end{bmatrix}, \quad B = \begin{bmatrix} 1 & 0 \\ 0 & 1 \end{bmatrix} = I.$$

Then $x(t) = \phi(t)[t\ 1]^T$ satisfies $Ax' + Bx = 0$ for any scalar function $\phi(t)$. This DAE is regular since $\det(\lambda A + B) \equiv 1$. However, it is not solvable since an initial value does not uniquely determine a solution. This example also fails to be solvable since solutions do not exist for all f because f must satisfy $[-1 \ t]^T f' = 0$ in order for a smooth solution to exist.

For linear constant coefficient systems, there were several equivalent definitions of the index. These definitions are not all equivalent for linear time varying DAE's.

Definition 2.4.3 *If* (2.4.1) *is a regular DAE, then the* local index *at* t, *denoted* $\nu_l(t)$, *is the index of the pencil* $\lambda A(t) + B(t)$.

Note that in (2.4.2) we have

$$x_n = (A_n + hB_n)^{-1} A_n x_{n-1} + (A_n + hB_n)^{-1} h f_n.$$

Theorem 2.3.4 is still valid, so that the local index will be important when discussing the condition of iteration matrices for the implicit methods of Chapters 3 and 4.

Recall that the index of a linear constant coefficient DAE was also defined as the number of iterations of the twofold process:

1. Using coordinate changes, reformulate the DAE to have explicit algebraic constraints.

2. Then differentiate the algebraic constraints until the system is reduced to an explicit ODE.

For linear time varying systems, this procedure requires the coefficient matrix of the derivative terms, namely $A(t)$, to have constant rank, a condition which does not in general hold true. When this procedure can be carried out for linear time varying DAE's, the number of iterations is still the index. However, there are solvable systems, with a well defined index, for which this reduction procedure cannot be carried out, even in theory. This can happen because solvability does not require the leading coefficient A to have either constant rank, or a smooth basis for its nullspace $\mathcal{N}(A)$.

Example 2.4.3. Let $\phi(t)$ be an infinitely differentiable function such that $\phi(t) > 0$ if $t > 0$ and $\phi(t) = 0$ if $t \leq 0$. Let $\psi(t)$ be an infinitely differentiable function such that $\psi(t) = 0$ if $t \geq 0$ and $\psi(t) > 0$ if $t < 0$. Then

$$\begin{bmatrix} 0 & \phi \\ \psi & 0 \end{bmatrix} x' + x = \begin{bmatrix} f_1 \\ f_2 \end{bmatrix}$$

is solvable on $(-\infty, \infty)$. Note that the pencil formed by the coefficients is nilpotent of index two if $t \neq 0$ and of index one if $t = 0$. The only solution of this DAE is

$$x = \begin{bmatrix} f_1 - \phi f_2' \\ f_2 - \psi f_1' \end{bmatrix}.$$

2.4. LINEAR TIME VARYING DAE'S

However, A has a rank change at $t = 0$ and $\mathcal{N}(A)$ changes from span$\{[0\ 1]^T\}$ to span$\{[1\ 0]^T\}$ as t goes from $t < 0$ to $t > 0$ so that no smooth coordinate changes P, Q defined on the whole real line $\mathcal{R}$ can be found which will make the system semi-explicit with explicit algebraic constraints.

This procedure of differentiating explicit constraints also shows that solutions of higher index linear time varying DAE's can involve derivatives of the coefficients as well as derivatives of the forcing function. This important fact will be discussed later in this section.

2.4.2 Structural Forms

We will now examine the behavior of linear time varying DAE's more carefully. For the linear constant coefficient case, the structure of the matrix pencil $\lambda A + B$ proved to be the key in Section 2.3. It will also be one key to understanding the behavior of numerical methods when applied to linear constant coefficient systems in the next chapter. Accordingly, we wish to develop the appropriate analogues of the pencil structure for the linear time varying case.

For the remainder of this section we consider linear time varying coordinate changes given by $x = Q(t)y$ and premultiplication by $P(t)$. The matrices P, Q are to be nonsingular on the interval of interest and at least as smooth as the coefficients of the DAE. In the proofs, one must often take Q to be slightly smoother, but we will not be discussing these technical points since they are not needed later. These changes of coordinates convert (2.4.1) to

$$PAQy' + (PAQ' + PBQ)y = Pf. \qquad (2.4.3)$$

We say that (2.4.1) and (2.4.3) are *analytically equivalent*. Since the index is a property of the DAE independent of coordinates, analytically equivalent DAE's have the same index (a proof is given later in this section). If Q is nonconstant, however, the local index sometimes changes under coordinate changes. For this reason, we will need to further restrict the P, Q in the later chapters when studying numerical methods. While we do not want to get into a detailed consideration of these results, the following facts will be useful.

Theorem 2.4.1 *Assume that (2.4.1) is a solvable DAE on the interval $\mathcal{I}$. Then the following are equivalent:*

1. *(2.4.1) is an implicit ODE*
2. $\nu = 0$
3. $\nu_l \equiv 0$
4. $\nu_l(t_0) = 0$ *for some $t_0 \in \mathcal{I}$*

where ν is the index and ν_l is the local index of (2.4.1).

Theorem 2.4.2 *Assume that* (2.4.1) *is solvable. Then* $\nu = 1$ *if and only if* $\nu_l \equiv 1$.

These two results show that ν and ν_l are the same for index zero and index one solvable systems. This is no longer true for higher index systems ($\nu \geq 2$) [50].

Theorem 2.4.3 *Assume that* (2.4.1) *is solvable on the interval* $\mathcal{I}$. *The following are preserved when transforming* (2.4.1) *to the analytically equivalent systems* (2.4.3):

1. $\nu_l \equiv 0$
2. $\nu_l \equiv 1$
3. $\nu_l(t) \geq 2$ *for all* $t \in \mathcal{I}$.

In fact, if $\nu_l \geq 2$ on $\mathcal{I}$, then it is possible to find an analytically equivalent system so that $\nu_l = 2$ on an open dense subset of $\mathcal{I}$ [122]. These results show that for many purposes we may consider three natural classes of solvable DAE's: implicit ODE's, index one, and index two.

The system (2.4.1) is semi-explicit if it is in the form

$$\begin{aligned} x_1' + B_{11}(t)x_1 + B_{12}(t)x_2 &= f_1 \\ B_{21}(t)x_1 + B_{22}(t)x_2 &= f_2. \end{aligned} \quad (2.4.4)$$

The semi-explicit DAE (2.4.4) has index one if and only if B_{22} is nonsingular for all t.

The semi-explicit index one DAE occurs frequently in applications. Many MOL problems are index one. The quadratic regulator control problem (1.3.4) is index one if the control weighting matrix R is nonsingular. The classical singular perturbation reduced order model (1.3.11) is usually assumed to be index one.

Many of the higher index semi-explicit DAE's arising in applications have a natural structure which we call Hessenberg form [76].

Definition 2.4.4 *The DAE* (2.4.1) *is in* Hessenberg form *of size* r *if it can be written as*

$$\begin{bmatrix} I & 0 & \cdot & \cdot & 0 \\ 0 & I & \cdot & \cdot & \cdot \\ \cdot & \cdot & I & \cdot & \cdot \\ \cdot & \cdot & \cdot & I & \cdot \\ 0 & \cdot & \cdot & \cdot & 0 \end{bmatrix} \begin{bmatrix} x_1' \\ \cdot \\ \cdot \\ \cdot \\ x_r' \end{bmatrix} + \begin{bmatrix} B_{11} & * & * & B_{1,r-1} & B_{1r} \\ B_{21} & * & * & B_{2,r-1} & 0 \\ 0 & * & * & * & \cdot \\ \cdot & \cdot & * & * & \cdot \\ 0 & \cdot & 0 & B_{r,r-1} & 0 \end{bmatrix} \begin{bmatrix} x_1 \\ \cdot \\ \cdot \\ \cdot \\ x_r \end{bmatrix} = \begin{bmatrix} f_1 \\ \cdot \\ \cdot \\ \cdot \\ f_r \end{bmatrix}$$

where x_i are vectors, B_{ij} are matrices, and the product $B_{r,r-1}B_{r-1,r-2}\cdots B_{1r}$ is nonsingular.

Proposition 2.4.1 *A DAE in Hessenberg form of size r is solvable and has index and local index r.*

2.4. LINEAR TIME VARYING DAE'S

The Hessenberg forms of size two and three are the most common. The Hessenberg form of size two is

$$x_1' + B_{11}x_1 + B_{12}x_2 = f_1$$
$$B_{21}x_1 = f_2$$

with $B_{21}B_{12}$ nonsingular. The Hessenberg form of size three is

$$x_1' + B_{11}x_1 + B_{12}x_2 + B_{13}x_3 = f_1$$
$$x_2' + B_{21}x_1 + B_{22}x_2 = f_2$$
$$B_{32}x_2 = f_3$$

where $B_{32}B_{21}B_{13}$ is nonsingular. Note that the $B_{i+1,i}$ will not, in general, be square much less invertible. It is only the product that needs to be square and nonsingular.

Many of the mechanics and variational problems as well as the trajectory control problem discussed in Section 1.3 will be shown in the next section to be nonlinear versions of the Hessenberg forms of size two and three. Some beam deflection problems are in Hessenberg form of size four [76]. The Hessenberg form will be important in our discussion of numerical methods in the following chapters.

For linear time varying DAE's, the (Kronecker) structure of the pencil $\lambda A + B$ is no longer as directly related to the solutions of the DAE as in the constant coefficient case. In Section 2.3 we saw that if N was a constant matrix and $N^k = 0$, $N^{k-1} \neq 0$, then $Nx' + x = f$ was solvable and had only one solution which involved $k-1$ derivatives of f. However, a solvable time varying system $Nx' + x = f$ with $N(t)^k \equiv 0$ may have more than one solution [50,56]. This occurs because in the linear time varying case it is possible to have $N^k \equiv 0$ but $(N\frac{d}{dt})^k \neq 0$. If $N(t)$ is structurally nilpotent, then we still have $(N\frac{d}{dt})^k = 0$ and a solution formula similar to (2.3.6) holds as with the linear constant coefficient case except that it also involves derivatives of N. Thus, if we want to have a general structure theory for linear time varying DAE's, we are naturally led to consider a form like (2.3.3), but with N structurally nilpotent. This form is sometimes called the standard canonical form [52].

Definition 2.4.5 *The system* (2.4.1) *is in* standard canonical form *if it is in the form*

$$\begin{bmatrix} I & 0 \\ 0 & N(t) \end{bmatrix} x' + \begin{bmatrix} C(t) & 0 \\ 0 & I \end{bmatrix} x = f(t)$$

where N is strictly lower (or upper) triangular.

Note that N need not have constant rank or index in this definition. If the reduction procedure of alternately isolating the constraint equations and then differentiating them can be used to show (2.4.1) is solvable and to find the index, then it can be shown that (2.4.1) is analytically equivalent to a system in standard canonical form. Example 2.4.3 demonstrates that not all solvable systems are analytically equivalent to one in standard canonical form. The situation is simpler if the coefficients are real analytic [65].

Theorem 2.4.4 *Suppose that A, B are real analytic. Then (2.4.1) is solvable if and only if it is analytically equivalent to a system in standard canonical form using real analytic coordinate changes.*

There is a weakened version of the standard canonical form that does hold for all solvable systems [58]. It is the most general result on the structure of solvable linear time varying DAE's known to us.

Theorem 2.4.5 *Suppose that (2.4.1) is solvable on the interval $\mathcal{I}$. Then it is analytically equivalent to*

$$\begin{bmatrix} I & G \\ 0 & N \end{bmatrix} z' + \begin{bmatrix} 0 & 0 \\ 0 & I \end{bmatrix} z = \begin{bmatrix} g \\ h \end{bmatrix}, \qquad (2.4.5)$$

where $Nz_2' + z_2 = h$ has only one solution for each function h. Furthermore, there exists a countable family of disjoint open intervals $\mathcal{I}_i$ such that $\bigcup \mathcal{I}_i$ is dense in $\mathcal{I}$ and on each $\mathcal{I}_i$, the system $Nz_2' + z_2 = h$ is analytically equivalent to one in standard canonical form of the form $Mw' + w = f$ with M structurally nilpotent.

This theorem shows that Example 2.4.3 is, in a sense, typical of how a DAE can be solvable but not equivalent to one in standard canonical form.

In applications, systems are not always easily viewed as one of the previously described forms. However, they can often be understood as combinations of these forms.

Definition 2.4.6 *The DAE (2.4.1) is an rth order (lower) triangular chain if it is in the form*

$$\begin{bmatrix} A_{11} & 0 & \cdot & 0 \\ A_{21} & A_{22} & \cdot & \cdot \\ * & * & * & 0 \\ A_{r1} & * & * & A_{rr} \end{bmatrix} \begin{bmatrix} x_1' \\ \cdot \\ \cdot \\ x_r' \end{bmatrix} + \begin{bmatrix} B_{11} & 0 & \cdot & 0 \\ B_{21} & B_{22} & \cdot & \cdot \\ * & * & * & 0 \\ B_{r1} & * & * & B_{rr} \end{bmatrix} \begin{bmatrix} x_1 \\ \cdot \\ \cdot \\ x_r \end{bmatrix} = \begin{bmatrix} f_1 \\ \cdot \\ \cdot \\ f_r \end{bmatrix}$$

and each subsystem $A_{ii}x_i' + B_{ii}x_i = f_i - \sum_{j=1}^{i-1}(A_{ij}x_j' + B_{ij}x_j)$ is either index one, index zero, in Hessenberg form, or in standard canonical form.

Any system which is analytically equivalent to a triangular chain is solvable. As an example, consider the quadratic regulator problem of Section 1.3 with control weighting matrix R singular, but $B^T RB$ nonsingular. Then the Euler-Lagrange equations (1.3.5) are not Hessenberg, but are equivalent to a triangular chain of a Hessenberg system of size three and an index one purely algebraic system [63]. More specific examples of these various forms will be given in the next section.

2.4.3 General Structure Theory

Many applications involving DAE's can be easily viewed as arising in one of the special forms or chains of these forms. Often, as will be seen in Chapters 3 and 4, this insures that certain numerical techniques such as backward differentiation formulas or certain implicit Runge-Kutta methods will work. However, this is not always the case. What is needed is a way to verify solvability and to develop general numerical methods which, although perhaps computationally more intensive, can be used when other methods fail.

The discussion so far in this section shows that for higher index DAE's, the solutions of (2.4.1) will depend not only on the derivatives of the forcing function f, but also on the derivatives of the coefficients. The following general results were developed in [56,58]. Here we shall use them to obtain characterizations of solvability and the consistent initial conditions. Their numerical implementation is discussed in [55,57].

Suppose that x is a solution of

$$A(t)x'(t) + B(t)x(t) = f(t).$$

For the remainder of this section let $c_i(t) = c^{(i)}(t)/i!$ for $i \geq 0$, $c = x, A, B, f$, that is, c_i is now the ith Taylor coefficient of c at time t rather than the ith component of the vector c. Then for each t, and j less than the smoothness of x, we have that

$$\begin{bmatrix} A_0 & 0 & \cdot & 0 \\ A_1 + B_0 & 2A_0 & \cdot & \cdot \\ * & * & * & 0 \\ A_{j-1} + B_{j-2} & 2A_{j-2} + B_{j-3} & * & jA_0 \end{bmatrix} \begin{bmatrix} x_1 \\ \cdot \\ \cdot \\ x_j \end{bmatrix} = - \begin{bmatrix} B_0 \\ \cdot \\ \cdot \\ B_{j-1} \end{bmatrix} x_0 + \begin{bmatrix} f_0 \\ \cdot \\ \cdot \\ f_{j-1} \end{bmatrix}$$

or

$$\mathcal{A}_j \mathbf{x}_j = -\mathcal{B}_j x_0 + \mathbf{f}_j. \qquad (2.4.6)$$

The *derivative array* $\mathcal{A}_j$ is singular for all t so that (2.4.6) does not uniquely determine $\mathbf{x}_j$ given x_0 and t. However, (2.4.6) could uniquely determine $x_1(t) = x'(t)$.

Definition 2.4.7 *The matrix $\mathcal{A}_j$ is smoothly 1-full if there is a smooth nonsingular $R(t)$ such that*

$$R(t)\mathcal{A}_j(t) = \begin{bmatrix} I & 0 \\ 0 & H(t) \end{bmatrix}$$

and the identity I is $m \times m$ where A, B are $m \times m$.

Given t and $x_0 = x(t)$, the systems of algebraic equations (2.4.6) will uniquely determine $x'(t)$ precisely when $\mathcal{A}_j$ is 1-full. Let $[R_{11}, \ldots, R_{1j}]$ be the first m rows of R. Then multiplying (2.4.6) by R and keeping only the first m rows yields

$$x_1 = -\sum_{i=1}^{j} R_{1i}B_{i-1}x_0 + \sum_{i=1}^{j} R_{1i}f_{i-1}$$

or

$$x'(t) = G(t)x(t) + g(t). \tag{2.4.7}$$

If $\mathcal{A}_j$ has constant rank, the R_{1i} will be as smooth as $\mathcal{A}_j$. This procedure has embedded the solutions of the DAE (2.4.1) into those of the ODE (2.4.7). As discussed further in [54,55], this approach allows one in principle to solve (2.4.1) numerically by applying ODE solvers to (2.4.7). Of more importance here is that this procedure provides a tool for understanding linear time varying and general DAE's. In particular, we can now give characterizations of solvability and the index.

Theorem 2.4.6 *Suppose that A, B are $2m$-times continuously differentiable and $Ax' + Bx = f$ is solvable on $\mathcal{I}$. Then*

1. *$\mathcal{A}_j$ has constant rank on $\mathcal{I}$ for $j = m+1$*
2. *$\mathcal{A}_j$ is smoothly 1-full on $\mathcal{I}$ for $j = m+1$*
3. *$\mathcal{R}(\mathcal{A}_j) + \mathcal{R}(\mathcal{B}_j) = \mathcal{R}^{mj}$ for every $t \in \mathcal{I}$, $1 \leq j \leq m+1$.*

This result is almost a characterization of solvability [58].

Theorem 2.4.7 *Suppose that A, B are $3m$-times continuously differentiable. Then (1)-(3) of Theorem 2.4.6 imply that $Ax' + Bx = f$ is solvable.*

The preceding results have several consequences. First, they provide a rigorous definition of the index for a solvable linear time varying DAE.

Proposition 2.4.2 *The index ν is the smallest integer such that $\mathcal{A}_{\nu+1}$ is 1-full and has constant rank.*

To show that the index is a property of the DAE and not the coordinate system, we need only show that the properties of 1-fullness and constant rank of the derivative array $\mathcal{A}_j$ are unaltered by analytic equivalence. To see this, let $\hat{\mathcal{A}}_j$ be the derivative array for $PAx' + PBx = Pf$, $\mathcal{A}_j$ the derivative array for $Ax' + Bx = f$, and $\tilde{\mathcal{A}}_j$ the derivative array for $A(Qy)' + B(Qy) = f$. Letting $P_i = P^{(i)}(t)/i!$, $Q_i = Q^{(i)}(t)/i!$, we have

$$\hat{\mathcal{A}}_j = \begin{bmatrix} P_0 & 0 & \cdot & 0 \\ P_1 & P_0 & \cdot & \cdot \\ * & * & * & 0 \\ P_{j-1} & * & * & P_0 \end{bmatrix} \mathcal{A}_j.$$

2.4. LINEAR TIME VARYING DAE'S

Thus $\tilde{\mathcal{A}}_j$ is 1-full and constant rank if and only if $\mathcal{A}_j$ is. Also, if $x = Qy$, then

$$\begin{bmatrix} x_0 \\ x_1 \\ \cdot \\ x_r \end{bmatrix} = \begin{bmatrix} Q_0 & 0 & \cdot & 0 \\ Q_1 & Q_0 & \cdot & \cdot \\ * & * & * & 0 \\ Q_r & * & * & Q_0 \end{bmatrix} \begin{bmatrix} y_0 \\ y_1 \\ \cdot \\ y_r \end{bmatrix}$$

so that $\tilde{\mathcal{A}}_j$ uniquely determines y_1 in terms of y_0, t if and only if $\mathcal{A}_j$ uniquely determines x_1 in terms of x_0, t.

This procedure also provides a characterization of the consistent initial conditions. A proof and discussion of the numerical implementation are in [57,60]. Recall that the *nullity* of a matrix is the dimension of its nullspace.

Theorem 2.4.8 *Let A, B be $2m$-times continuously differentiable and assume that (2.4.1) is solvable. Let j be such that $\mathcal{A}_j$ is 1-full and constant rank. Then the linear manifold of consistent initial conditions is precisely the set of all $x(t_0)$ such that $\mathcal{A}_j(t_0)\mathbf{x}_j = \mathbf{f}_j(t_0) - \mathcal{B}_j(t_0)x(t_0)$ is consistent, equivalently, $\mathbf{f}_j - \mathcal{B}_j x(t_0) \in \mathcal{R}(\mathcal{A}_j(t_0))$. Thus one may determine the manifold of consistent initial conditions by*

1. *Compute W such that $\mathcal{A}_j W = 0$ and rank(W) = nullity($\mathcal{A}_j$).*
2. *Solve $W^T \mathcal{B}_j x(t_0) = W^T \mathbf{f}_j$ for $x(t_0)$.*

Another important consequence of this approach is that it proves the continuity of solutions with respect to variations of consistent initial conditions within the manifold of consistent initial conditions. It also gives insight on the solutions as functions of the coefficients. The two key observations are that G in (2.4.7) involves ν derivatives of A, B and G will be a continuous function of these derivatives as long as $\mathcal{A}_j$ stays 1-full with constant rank and solvability is maintained [58].

Many approximation schemes and engineering design algorithms are based on the fact that the solutions of the linear constant coefficient system

$$x'(t) = G(\hat{t})x(t) + g(t)$$

approximate the solutions of

$$x'(t) = G(t)x(t) + g(t)$$

for t near $\hat{t}$. The dependence of G on the derivatives of A, B means that, in general, the linear constant coefficient DAE

$$A(\hat{t})x'(t) + B(\hat{t})x(t) = f(t) \qquad (2.4.8)$$

need not be a good approximation to the linear time varying DAE

$$A(t)x'(t) + B(t)x(t) = f(t) \qquad (2.4.9)$$

even for t arbitrarily close to $\hat{t}$. In fact, the systems (2.4.8) and (2.4.9) do not have to be both solvable, and if solvable do not have to have the same index nor the same dimensional manifold of consistent initial conditions.

2.5 Nonlinear Systems

2.5.1 Solvability and the Index

Many of the most important DAE's are nonlinear. The theory just presented for linear time varying DAE's is useful as a guide in the derivation of an analogous theory still under development for nonlinear systems. The nonlinearity, of course, will make most results local in nature. Also, as it will be much more difficult to establish necessary and sufficient conditions, we will usually have to settle for sufficient conditions.

First, we wish to elaborate on the definition of solvability and the intuitive definition of the index given in Section 2.2 for the general nonlinear DAE

$$F(t, y, y') = 0. \tag{2.5.1}$$

Following the development of the linear time varying case in the last section, we consider the system of equations

$$\begin{aligned} F(t, y, y') &= 0 \\ \frac{d}{dt} F(t, y, y') &= 0 \\ &\vdots \\ \frac{d^{j-1}}{dt^{j-1}} F(t, y, y') &= 0 \end{aligned} \tag{2.5.2}$$

which can be written as

$$\begin{aligned} F_{[0]}(t, y, y') &= 0 \\ F_{[1]}(t, y, y', y'') &= 0 \\ &\vdots \\ F_{[j-1]}(t, y, y', \dots, y^{(j-1)}, y^{(j)}) &= 0 \end{aligned} \tag{2.5.3}$$

where, for example, $F_{[1]} = F_{y'} y'' + F_y y' + F_t$. We shall write (2.5.3) as

$$\mathbf{F}_j(t, y, \mathbf{y}_j) = 0 \tag{2.5.4}$$

where, as in the last section,

$$\mathbf{y}_j = \begin{bmatrix} y' \\ \vdots \\ y^{(j)} \end{bmatrix}.$$

The system (2.5.4) is the nonlinear analogue of the system (2.4.6) for linear time varying DAE's. In this notation we may express the definition of the index as

2.5. NONLINEAR SYSTEMS

Definition 2.5.1 *The* index ν *of* (2.5.1) *is the smallest* ν *such that* $\mathbf{F}_{\nu+1}$ *uniquely determines the variable* y' *as a continuous function of* y, t.

From [59] we have

Proposition 2.5.1 *Sufficient conditions for* (2.5.4) *to uniquely determine* y' *as a continuous function of* y, t *are that the Jacobian matrix of* $\mathbf{F}_j$ *with respect to* $\mathbf{y}_j$, *denoted* $\partial \mathbf{F}_j / \partial \mathbf{y}_j$, *is 1-full with constant rank and* (2.5.4) *is consistent.*

For the linear time varying DAE, this Jacobian matrix is precisely the matrix $\mathcal{A}_j$ of the last section. Again one can show that the properties of 1-full and constant rank of the Jacobian matrix are preserved by nonlinear coordinate changes of premultiplication by $P(y,t)$ and letting $y = Q(x,t)$ with Q_x nonsingular [59].

As with the definition of solvability in Section 2.2, all of these statements, including the preceding definition and proposition, are taken to hold locally on open subsets of $\mathcal{R}^{jm+1}$. For example, consider the purely algebraic DAE $y^2 = t$ so that $\mathbf{F}_1 = [y^2 - t, 2yy' - 1]^T$. The DAE $y^2 = t$ is index one on the open sets in $\mathcal{R}^3$ of the form $(t, y, y') \in \mathcal{R} \times (0, \infty) \times (0, \infty)$ or $(t, y, y') \in \mathcal{R} \times (-\infty, 0) \times (0, \infty)$.

An alternative, but less general, characterization of the index is the number of iterations of the following (theoretical) procedure needed to convert the DAE into an ODE:

1. If $F_{y'}$ is nonsingular, then stop.

2. Suppose that $F_{y'}$ has constant rank and that nonlinear coordinate changes make (2.5.1) semi-explicit. Differentiate the constraint equation, let $F = 0$ denote the new DAE, and return to (1).

As with the linear time varying case, it is not possible to carry out this procedure even theoretically for all solvable DAE's. However, in some examples and techniques it provides an easy way to verify assertions about the index.

The nonlinear form of the implicit Euler method with stepsize h applied to (2.5.1) is

$$F\left(t_n, y_n, \frac{y_n - y_{n-1}}{h}\right) = 0, \qquad (2.5.5)$$

which will need to be solved by a nonlinear equation solver. The Jacobian of F in (2.5.5) with respect to y_n is $(1/h)F_{y'} + F_y$ so that this pencil will be important. For the linear case, this was the pencil $(1/h)A + B$.

Definition 2.5.2 *The* local index ν_l *of* (2.5.1) *at* $(\hat{t}, \hat{y}, \hat{y}')$ *is the index of the pencil* $\lambda F_{y'}(\hat{t}, \hat{y}, \hat{y}') + F_y(\hat{t}, \hat{y}, \hat{y}')$.

Proposition 2.5.2 *Suppose that* $F_{y'}$ *has constant rank. Then* $\nu = 1$ *if and only if* $\nu_l \equiv 1$.

Finally, note that the semi-explicit DAE

$$\begin{aligned} x_1' &= F_1(x_1, x_2, t) \\ 0 &= F_2(x_1, x_2, t) \end{aligned}$$

is index one if and only if $\partial F_2/\partial x_2$ is nonsingular.

2.5.2 Structural Forms

With nonlinear systems it is especially important to have recognizable forms with known properties. There are nonlinear versions of all the forms of the last section.

Definition 2.5.3 *The DAE (2.5.1) is in* Hessenberg form of size r *if it is written*

$$\begin{aligned} x_1' &= F_1(x_1, x_2, \ldots, x_r, t) \\ x_2' &= F_2(x_1, x_2, \ldots, x_{r-1}, t) \\ &\vdots \\ x_i' &= F_i(x_{i-1}, x_i, \ldots, x_{r-1}, t), \quad 3 \le i \le r-1 \\ &\vdots \\ 0 &= F_r(x_{r-1}, t) \end{aligned}$$

and $(\partial F_r/\partial x_{r-1})(\partial F_{r-1}/\partial x_{r-2})\cdots(\partial F_2/\partial x_1)(\partial F_1/\partial x_r)$ *is nonsingular.*

The Hessenberg form of size two is

$$\begin{aligned} x_1' &= F_1(x_1, x_2, t) \\ 0 &= F_2(x_1, t) \end{aligned}$$

with $(\partial F_2/\partial x_1)(\partial F_1/\partial x_2)$ nonsingular. The Hessenberg form of size three is

$$\begin{aligned} x_1' &= F_1(x_1, x_2, x_3, t) \\ x_2' &= F_2(x_1, x_2, t) \\ 0 &= F_3(x_2, t) \end{aligned} \quad (2.5.6)$$

with $(\partial F_3/\partial x_2)(\partial F_2/\partial x_1)(\partial F_1/\partial x_3)$ nonsingular. Assuming the F_i are sufficiently differentiable, the Hessenberg form of size r is solvable and has index and local index r.

Definition 2.5.4 *The DAE (2.5.1) is in* standard canonical form of size r *if it can be written as*

$$\begin{aligned} x_1' &= F_1(x_1, t) \\ N(x_1, x_2, t)x_2' &= x_2 + h(x_1, t) \end{aligned}$$

where N is an $r \times r$ block strictly lower triangular matrix.

2.5. NONLINEAR SYSTEMS

Finally, we have

Definition 2.5.5 *The DAE (2.5.1) is a (lower) triangular chain of order r, if it can be written as*

$$\begin{aligned} F_1(x_1', x_1, t) &= 0 \\ F_2(x_1', x_2', x_1, x_2, t) &= 0 \\ &\vdots \\ F_r(x_1', \ldots, x_r', x_1, \ldots, x_r, t) &= 0 \end{aligned}$$

and the system defined by the function F_i is either an implicit ODE, index one, a Hessenberg form, or in standard canonical form in the variable x_i.

A triangular chain is always solvable. To illustrate how these forms occur, we shall briefly consider some of the applications given in Section 1.3.

The constrained variational problems (1.3.1) are in Hessenberg form of size three since they are in the form

$$\begin{aligned} H(x_1, x_2)x_1' &= F_1(x_1, x_2, x_3, t) \\ x_2' &= F_2(x_1, x_2, t) \\ 0 &= F_3(x_2) \end{aligned}$$

with $x_1 = u$, $x_2 = x$, $x_3 = \lambda$.

The optimal control problem (1.3.2),(1.3.3) is semi-explicit index one if g_{uu} is nonsingular. If g_{uu} is singular, but $f_x^T g_{uu} f_x$ is nonsingular, the index is three. If both these matrices are singular, the index is at least five.

We will see in Chapter 6 that the prescribed path control problem (1.3.6), (1.3.8) is in Hessenberg form of size three. The robotics problem in [90] is in Hessenberg form of size five.

As a final example, consider the chemical reactor model (1.3.9). If we treat T_c as known, the resulting DAE (1.3.9a)-(1.3.9c) is semi-explicit index one in T, R, C. If, however, we consider T_c to be an unknown to be determined, we get the DAE (1.3.9a)-(1.3.9d) in T, T_c, C, R. This system is not in Hessenberg form as written, but it can be written as

$$\begin{aligned} C' &= f_1(C, R) \\ 0 &= f_2(C, t) \\ \hline T' &= f_3(T, T_c, R, C) \\ 0 &= f_4(T, R, C). \end{aligned} \tag{2.5.7}$$

Letting

$$x_1 = \begin{bmatrix} C \\ R \end{bmatrix}, \quad x_2 = \begin{bmatrix} T \\ T_c \end{bmatrix}$$

we have that (2.5.7) is

$$F_1(x_1', x_1, t) = 0 \qquad (2.5.8a)$$
$$F_2(x_2', x_2, x_1) = 0 \qquad (2.5.8b)$$

where (2.5.8a) and (2.5.8b) are in Hessenberg form of size two in x_1 and x_2, respectively. Thus (2.5.7) is a lower triangular chain of size two.

2.5.3 Index Reduction and Constraint Stabilization

From the point of view of the numerical solution, it is desirable for the DAE to have an index which is as small as possible. As we have seen, a reduction of the index can be achieved by differentiating the constraints. However, one of the difficulties with reducing the index in this way is that the numerical solution of the resulting system need no longer satisfy the original constraints exactly. This can have serious implications for some problems where, for example, the constraints reflect important physical properties. Thus it is quite useful to have a means of introducing constraints which have been lost, through differentiations or other manipulations, back into the system in such a way that the resulting system has a structure which is computationally tractable. Such an idea was proposed by Gear [118] and has several important applications.

The framework for this idea is as follows. Suppose that we have the DAE (2.5.1) and through a series of differentiations and coordinate changes we arrive at the ODE

$$y' = f(y, t) \qquad (2.5.9a)$$

and along the way there were the constraints

$$0 = g(y, t). \qquad (2.5.9b)$$

We note that these constraints might also be present in the system because they represent invariants of the solution of (2.5.9a) such as, for example, conservation of energy [117]. The relevant property is that the analytic solution to (2.5.9a) satisfies the constraints (2.5.9b), whereas the numerical solution may not, and this is what we are trying to correct. Suppose also that g_y has full row rank – that is, the constraints (2.5.9b) are linearly independent. Let μ be a new variable of size rank(g_y) and consider the semi-explicit DAE

$$y' = f(y, t) + g_y^T \mu \qquad (2.5.10a)$$
$$0 = g(y, t). \qquad (2.5.10b)$$

Theorem 2.5.1 *Assume that g_y has full row rank and that (2.5.9b) is a characterization of the solution manifold for (2.5.1). More precisely, there is a solution to (2.5.1) satisfying $y(\hat{t}) = \hat{y}$ if and only if $g(\hat{y}, \hat{t}) = 0$ for fixed $\hat{t}$. Then, the only solutions of (2.5.10) are $\mu = 0$ and y a solution of (2.5.9).*

2.5. NONLINEAR SYSTEMS

Proof. Since this theorem will be useful later and its verification helps to understand the index we shall give the proof. Clearly if y is a solution of (2.5.9) and $\mu = 0$, then y, μ is a solution of (2.5.9). Suppose then that (2.5.9b) characterizes the solution manifold and that g_y has full row rank. Let $(y(t), t)$ be such that $g(y, t) = 0$. Then (2.5.10) has solutions y, μ. Differentiation of (2.5.9b) with respect to t gives

$$g_t(y,t) + g_y(y,t)y' = 0 \tag{2.5.11}$$

and hence by (2.5.9a) we have that $g = 0$ implies

$$g_t + g_y f = 0. \tag{2.5.12}$$

Multiplying (2.5.10a) by g_y and using (2.5.11), (2.5.12) gives that $g_y g_y^T \mu = 0$ and hence, by the full rank assumption on g_y, that $\mu = 0.\square$

Knowing that $g = 0$ completely characterizes the solution manifold is not always possible. In fact, these equations may arise because of known physical constraints rather than differentiations. Even in this situation, the system (2.5.10) can still sometimes be useful for an over determined system like (2.5.9).

Theorem 2.5.2 *Assume that g_y has full row rank. Then the semi-explicit DAE (2.5.10) is index two. Also, the solutions of (2.5.10) include y a solution of (2.5.9), and $\mu = 0$.*

Proof. A calculation like that of the proof of the previous theorem gives, after differentiation of (2.5.10b) that

$$g_t = -g_y f - g_y g_y^T \mu. \tag{2.5.13}$$

But g is independent of μ. Thus the full rank assumption implies that (2.5.13) determines μ in terms of y, t. A second differentiation gives μ' and thus the index is two.$\square$

As a more specific example, consider the Euler-Lagrange formulation of the constrained mechanical system from Section 1.3,

$$\begin{aligned} x' &= u \\ \frac{d}{dt}\frac{\partial}{\partial u}T(x,u) &= T_x + f(x,u,t) + G^T\lambda \\ 0 &= \phi(x) \end{aligned}$$

with $G = \phi_x$ and λ the Lagrange multiplier. Assume that T_{uu} is a positive definite matrix. Then this system is an index three Hessenberg system which may be written

$$\begin{aligned} x' &= u & (2.5.14a) \\ Mu' &= g(x,u,t) + G^T\lambda & (2.5.14b) \\ 0 &= \phi(x) & (2.5.14c) \end{aligned}$$

where $g(x,u,t) = T_x + f - T_{xu}u$. If (2.5.14c) is differentiated and the approach of the previous two theorems is used, we obtain the semi-explicit index two DAE

$$\begin{aligned} x' &= u + G^T\mu \\ Mu' &= g(x,u,t) + G^T\lambda \\ 0 &= Gu \\ 0 &= \phi(x) \end{aligned}$$

and (x,u,λ,μ) is a solution of this DAE if and only if $\mu = 0$ and (x,u,λ) is a solution of (2.5.14).

Another way that the index can sometimes be lowered without performing any explicit differentiations is also pointed out in [118]. To illustrate, consider the index three linear constant coefficient DAE given in semi-explicit form

$$\begin{aligned} x_2' &= x_1 + f_1(t) \\ x_3' &= x_2 + f_2(t) \\ 0 &= x_3 + f_3(t). \end{aligned} \quad (2.5.15)$$

Introduce the new variable $z_1' = x_1$. Then (2.5.15) becomes

$$\begin{aligned} x_2' - z_1' &= f_1 \\ x_3' &= x_2 + f_2 \\ 0 &= x_3 + f_3 \end{aligned} \quad (2.5.16)$$

which is an index two DAE in (z_1, x_2, x_3). This index reduction option is available any time we have a DAE with variables, such as x_1 in this example, whose derivatives do not explicitly appear in the DAE. In more complicated systems with more variables, the substitution may not reduce the index. Intuitively, one can think of the *index of a variable* as the number of differentiations needed to find its derivative. This substitution may reduce the index of only some of the variables. The index of a variable is closely related to the concept of dynamic order in nonlinear control theory.

If we are actually interested in x_1, then this may not be very helpful numerically since after solving (2.5.16) for z_1, we still need to differentiate the computed z_1 to get x_1. However, if we are only interested in the other variables, such as x_2, x_3 in (2.5.16), then this method can sometimes reduce the index of the system we need to solve. For example, in the variational problem (2.5.14), we could replace λ by ψ' to get an index two DAE in (x, u, ψ).

This type of index reduction does suffer from the fact that the lower index problem has more solutions, just as differentiating the constraints would. However, no actual differentiations need be done for the reduction and the system size is not increased by introducing a new variable μ so that it is simpler to carry out. This approach is also very useful in understanding the numerical

2.5. NONLINEAR SYSTEMS

methods of the next chapter. In particular, it enables us to make assertions about behavior for higher index variables based on results for the methods on lower index problems.

There is a relationship between the index of semi-explicit systems and general systems that is worth stating as a "rule of thumb" and then examining more carefully in the chapters to follow.

Rule of Thumb: *The semi-explicit case is much like the general case of one lower index.*

This 'rule' can be motivated in several ways, with a linear time varying argument given in [53]. Here we proceed as follows. We have already seen how to reduce the index of a semi-explicit DAE (1.1.3) by substitutions of the form $u' = y$, thereby creating a fully-implicit DAE. The next proposition provides the relationship in the reverse direction.

Proposition 2.5.3 *Suppose that $F(t, y, y') = 0$ is a solvable DAE of index ν. Then*

$$\begin{aligned} y' &= z \\ 0 &= F(t, y, z) \end{aligned}$$

is a semi-explicit DAE of index $\nu + 1$.

Proof. Observe that we already have y' and know that k differentiations will determine z in terms of y, t. Then one more differentiation will determine z'. □

A numerical method may be applied to either the original DAE or to the enlarged system, but it is important to note that the resulting convergence and stability properties of the schemes may be quite different because of the change in the index. Until recently, the selection of the DAE form was often overlooked in practice when applying a numerical method. März [181] was the first to seriously study general numerical methods when applied to both the original and the enlarged DAE system. In general, we expect numerical schemes to perform better when applied to the lower index system. However, this is only true if the lower index system has similar stability properties to the original DAE [108].

Chapter 3

Multistep Methods

3.1 Introduction

In this chapter we study the convergence, order and stability properties of linear multistep methods (LMM's) applied to DAE's. We take an historical perspective in the beginning, to motivate why the backward differentiation formulas (BDF) have emerged as the most popular and hence best understood class of linear multistep methods for general DAE's.

The first general technique for the numerical solution of DAE's, proposed in 1971 by Gear in a well-known and often cited paper [113], utilized the BDF. This method was initially defined for systems of differential equations coupled to algebraic equations

$$\begin{aligned} x' &= f(x,y,t) \\ 0 &= g(x,y,t), \end{aligned} \tag{3.1.1}$$

where y is a vector of the same dimension as g. The algebraic variables y are treated in the same way as the differential variables x for BDF, and the method was soon extended to apply to any fully-implicit DAE system

$$F(t,y,y') = 0. \tag{3.1.2}$$

The simplest first order BDF method is the implicit Euler method, which consists of replacing the derivative in (3.1.2) by a backward difference

$$F\left(t_n, y_n, \frac{y_n - y_{n-1}}{h}\right) = 0,$$

where $h = t_n - t_{n-1}$. The resulting system of nonlinear equations for y_n at each time step is then usually solved by Newton's method. The k-step (constant-stepsize) BDF consists of replacing y' by the derivative of the polynomial which interpolates the computed solution at $k+1$ times $t_n, t_{n-1}, \ldots, t_{n-k}$, evaluated at t_n. This yields

$$F\left(t_n, y_n, \frac{\rho y_n}{h}\right) = 0, \tag{3.1.3}$$

where $\rho y_n = \sum_{i=0}^{k} \alpha_i y_{n-i}$ and α_i, $i = 0, 1, \ldots, k$ are the coefficients of the BDF method. The k-step BDF method is stable for ODE's for $k < 7$. An introduction to the properties of multistep methods, and in particular BDF, for ODE's can be found in [114,137,159].

Implicit in the application of numerical ODE methods such as multistep or Runge-Kutta to (3.1.1) and (3.1.2) are several assumptions. First, the problem must make sense mathematically. That is, the DAE system must be *solvable* in the sense of Section 2.2. Second, the method must be *implementable*. In other words, the method and problem must be such that the nonlinear system of equations which must be solved on each time step has a solution. For the k-step BDF method, this is the system (3.1.3).

Sometimes it is obvious that the method is implementable. In general, showing implementability consists of two parts. First it is shown, or assumed, that (3.1.3) is consistent. Then it is argued that for small enough h, it is possible to recursively find the y_n. A key role is played by the nonsingularity of the Jacobian of (3.1.3) and the assumption that the original DAE (3.1.2) is solvable so that the implicit function theorem can be utilized.

Following the paper by Gear [113], several early codes implementing the BDF methods were written [27,35,216]. Multistep methods other than BDF have also been considered in the DAE literature. Söderlind [233] developed and analyzed methods for the semi-explicit index one system (3.1.1), where the system was partitioned into a stiff system which includes the algebraic equations, and a nonstiff subsystem. The nonstiff subsystem was solved by the classical fourth order explicit Runge-Kutta method, while the stiff/algebraic subsystem was solved by the three-step BDF. This type of approach has the advantage of treating part of the system explicitly, and the disadvantage that the efficiency and robustness of the method depends on the quality of and the ability to make the partitioning. Liniger [164] considered a two-step one-leg formula for index one DAE's where the algebraic constraints are clearly identifiable. This method has stronger stability properties than the BDF for problems which are very nonlinear or which require frequent stepsize changes. The most general and extensive work on convergence analysis for linear multistep methods and their one-leg twins has been conducted by Griepentrog and März, and reported in the monograph [132]. These results are directed primarily at index one systems and will be discussed in Section 3.4. Still, for general index one DAE's, the greatest amount of success in solving problems from applications has thus far been attained by codes based on BDF. As we will see later in this chapter, much of the success of BDF codes has undoubtedly been due to the extraordinary stability and accuracy properties of BDF applied to DAE's, including even many higher index DAE's.

A second generation of BDF implementations began to emerge in the early 1980's, along with a growing recognition of the importance of DAE's in many scientific and engineering applications. These are the codes DASSL [201] and LSODI [143]. The DASSL code is described extensively in Chapter 5. Many

3.1. INTRODUCTION

DAE problems were successfully solved using these codes, thereby encouraging the BDF approach. Still, not all DAE's were solved successfully with the BDF codes, and in the early 1980's research papers began reporting these difficulties [200,229]. To illustrate one of the first problems observed with the BDF codes, which is due to the possibility of formulating DAE systems (3.1.1) or (3.1.2) with an index greater than one, we will analyze a simple example. One of the interesting features of this problem is that while the constant stepsize BDF formulas converge to the correct order of accuracy, the BDF codes fail. Some of the difficulties are that the solution is not accurate for the first two steps, the formulas (particularly the implicit Euler method) are not accurate when the stepsize changes, and the error estimates used in the codes to control the stepsize are not realistic for this type of problem.

Consider the index three system

$$\begin{aligned} x_1' &= x_2 \\ x_2' &= x_3 \\ 0 &= x_1 - g(t). \end{aligned} \qquad (3.1.4)$$

The exact solution to the DAE is clearly $x_1(t) = g(t)$, $x_2(t) = g'(t)$, $x_3(t) = g''(t)$. Discretize (3.1.4) by the implicit Euler method to obtain

$$\begin{aligned} x_{1,n} &= g(t_n) \\ x_{2,n} &= (x_{1,n} - x_{1,n-1})/h \\ x_{3,n} &= (x_{2,n} - x_{2,n-1})/h. \end{aligned}$$

Ignoring round-off error, the value of x_1 will be determined exactly on all steps, even if the initial value $x_{1,0}$ is wrong. Now suppose the initial values given at t_0 are inconsistent (that is, they do not satisfy (3.1.4)). Then the values $x_{2,1}$ and $x_{3,1}$ obtained at the end of the first step will be incorrect. After two steps, $x_{2,2}$ will be $O(h)$ accurate since it is determined by the first divided difference of $g(t)$. However, $x_{3,2}$ is still incorrect. On the third step the value for $x_{3,3}$ is $O(h)$ accurate because it is obtained by the second divided difference of $g(t)$. Therefore, in spite of incorrect initial values, after three steps of length h with the implicit Euler method, the numerical solution is $O(h)$ accurate. (If consistent initial values are given, the solution is $O(h)$ accurate at the end of the second step.) A convergence result for BDF applied to constant coefficient linear DAE systems was first given by Sincovec et al.[229] in 1981. For completeness, and to illustrate some of the ideas usually employed in convergence analysis for more complicated systems of DAE's, we give the argument here. It is important to note that for higher index ($\nu \geq 2$) systems, the numerical solution converges in an interval *bounded away from the initial time*.

Theorem 3.1.1 *The k-step constant stepsize BDF method ($k < 7$) applied to constant coefficient linear DAE systems of index ν is convergent of order $O(h^k)$ after $(\nu - 1)k + 1$ steps.*

Proof. Consider the following linear constant coefficient system

$$Ax' + Bx = f(t), \qquad (3.1.5)$$

where A, B are constant square matrices. Discretize (3.1.5) with the k-step BDF to obtain

$$\frac{1}{h}\sum_{j=0}^{k}\alpha_j A x_{n-j} + B x_n = f_n, \qquad (3.1.6)$$

where $f_n = f(t_n)$. Let

$$\rho x_n = \sum_{j=0}^{k}\alpha_j x_{n-j}.$$

Then (3.1.6) can be written simply as

$$A\left(\frac{\rho x_n}{h}\right) + B x_n = f_n. \qquad (3.1.7)$$

From Theorem 2.3.2 we know that if the matrix pencil $\lambda A + B$ is regular, there exist nonsingular matrices P, Q such that

$$PAQ = \begin{bmatrix} I & 0 \\ 0 & N \end{bmatrix}, \quad PBQ = \begin{bmatrix} C & 0 \\ 0 & I \end{bmatrix}, \qquad (3.1.8)$$

where N is a matrix of nilpotency ν and I is an identity matrix. Let $x_n = Q y_n$ and premultiply the discretized system (3.1.7) by the nonsingular matrix P to obtain

$$PAQ\left(\frac{\rho y_n}{h}\right) + PBQ y_n = P f_n, \qquad (3.1.9)$$

where we use the fact that $Q\rho = \rho Q$. Using (3.1.8) and letting $P f_n = g_n$, system (3.1.9) can be written as two uncoupled subsystems

$$\frac{\rho y_{1,n}}{h} + C y_{1,n} = g_{1,n} \qquad (3.1.10a)$$

$$N\left(\frac{\rho y_{2,n}}{h}\right) + y_{2,n} = g_{2,n}. \qquad (3.1.10b)$$

The first equation (3.1.10a) is just the difference equation obtained by applying the BDF method to the ODE $y_1' + C y_1 = g_1(t)$. It is well-known that the numerical solution of an explicit ODE by the k-step BDF is $O(h^k)$ accurate. The second equation (3.1.10b) represents the BDF method applied to the purely algebraic subsystem $N y_2' + y_2 = g_2(t)$. Clearly we need only analyze the convergence of the numerical solution to the subsystem (3.1.10b). For simplicity, we drop the subscript on y_2 and consider

$$N\left(\frac{\rho y_n}{h}\right) + y_n = g_n. \qquad (3.1.11)$$

3.1. INTRODUCTION

In Chapter 2 we derived the analytic solution to (3.1.11)

$$y(t) = \sum_{i=0}^{\nu-1}(-1)^i N^i g^{(i)}(t), \qquad (3.1.12)$$

where $g^{(i)} = d^i g/dt^i$. Solving (3.1.11) for y_n, and noting that

$$\left(N\left(\frac{\rho}{h}\right) + I\right)^{-1} = \sum_{i=0}^{\nu-1}(-1)^i N^i \left(\frac{\rho}{h}\right)^i,$$

we obtain

$$y_n = \sum_{i=0}^{\nu-1}(-1)^i N^i \left(\frac{\rho}{h}\right)^i g_n.$$

Since for the BDF coefficients

$$\left(\frac{\rho}{h}\right)^i g_n = g_n^{(i)} + O(h^k),$$

it follows that

$$y_n = \sum_{i=0}^{\nu-1}(-1)^i N^i \left(g_n^{(i)} + O(h^k)\right).$$

Note that the numerical solution y_n does not depend on the initial value y_0, and in fact depends only on the values of g at the last $(\nu - 1)k + 1$ points. Comparing the numerical solution y_n to the analytic solution $y(t_n)$ given in (3.1.12), we see that the local and global errors are both $O(h^k)$. In contrast, for BDF applied to an explicit ODE, the local error is one order of accuracy higher than the global error. □

For variable stepsize BDF methods, Gear et al.[119] showed that if the ratio of the adjacent steps is kept bounded, then the global error in the numerical solution for k-step BDF applied to index ν constant coefficient systems is $O(h_{\max}^q)$, where $q = \min(k, k - \nu + 2)$. Therefore, the numerical solution of an index three system by the implicit Euler method using variable stepsizes is expected to contain $O(1)$ errors! This fact is important since all current general purpose DAE codes based on the BDF start off the integration process with the first order formula. As variable step methods, they will fail to integrate even a simple linear constant coefficient index three system. Of course, if the index of the system is known, one might design a code which starts with an appropriately high order formula.

A second difficulty which arises for the BDF method, and for numerical ODE methods in general, concerns stability for general higher index DAE systems. There are index two and three DAE systems for which the implicit Euler method, and in fact all of the multistep and Runge-Kutta methods, are unstable [122]. For example, consider the following solvable linear index two DAE

$$\begin{bmatrix} 0 & 0 \\ 1 & \eta t \end{bmatrix} x' + \begin{bmatrix} 1 & \eta t \\ 0 & 1+\eta \end{bmatrix} x = \begin{bmatrix} g(t) \\ 0 \end{bmatrix}, \quad (3.1.13)$$

which has the solution $x_1(t) = g(t) + \eta t g'(t)$, $x_2(t) = -g'(t)$. Note that this DAE does not even have a regular matrix pencil if $\eta = -1$. The difference equations obtained by discretizing (3.1.13) by the implicit Euler method are

$$\begin{aligned} x_{2,n} &= \frac{\eta}{1+\eta} x_{2,n-1} - \frac{g(t_n) - g(t_{n-1})}{(1+\eta)h} \\ x_{1,n} &= g(t_n) - \eta t_n x_{2,n}. \end{aligned} \quad (3.1.14)$$

These recurrence relations are unstable when $\eta < -.5$.

In Section 3.2 we give the convergence results for BDF methods. Because multistep and Runge-Kutta methods are not stable and convergent for all higher index DAE systems, research has focused on obtaining convergence results for classes of problems which arise commonly in applications. These are primarily semi-explicit index one systems, fully-implicit index one systems, semi-explicit index two systems, and index three systems in Hessenberg (see Section 2.5.2) form. In Subsection 3.2.1 we begin with the simple situation of semi-explicit index one systems. In Subsection 3.2.2 we examine fully-implicit nonlinear index one systems, deriving the main convergence result for BDF methods. This result is particularly important because BDF are the methods used in general purpose codes such as DASSL [201] and LSODI [143], which are directed primarily at fully-implicit index one systems. In Subsection 3.2.3 we turn our attention to semi-explicit index two systems, where we show that constant stepsize k-step BDF methods converge to order $O(h^k)$ under appropriate assumptions. In Subsection 3.2.4 we discuss the results which have been obtained for BDF methods applied to index three systems in Hessenberg form.

In Section 3.3 we address the relationship between convergence results which have been obtained for BDF methods applied to stiff ODE's and the results which have been given for DAE's.

The approach taken by Griepentrog and März [132] for the analysis of multistep methods is different from the one we have taken in this monograph both in its details and its philosophy. However, it is important because of its generality and the insight it gives into the development and analysis of methods for the numerical solution of DAEs. Therefore we have deferred a discussion of these results to the last section of this chapter, where we outline some of the main ideas and results related to multistep methods.

3.2 BDF Convergence

3.2.1 Semi-Explicit Index One Systems

Semi-explicit index one systems are an important subset of index one systems which arise frequently in problems in science and engineering. Recall that these systems are written as

$$x' = f(x,y,t) \qquad (3.2.1a)$$
$$0 = g(x,y,t), \qquad (3.2.1b)$$

where $(\partial g/\partial y)^{-1}$ exists and is bounded in a neighborhood of the exact solution.

The application of linear multistep methods to these systems is so straightforward that we will describe it for the general case, rather than confining ourselves to BDF as is done in the remainder of this section. The simplest and most natural implementation of linear multistep methods applied to (3.2.1) is to require the constraint to be satisfied at each step,

$$\sum_{j=0}^{k} a_j x_{n-j} = h \sum_{j=0}^{k} b_j f(x_{n-j}, y_{n-j}, t_{n-j}) \qquad (3.2.2a)$$
$$0 = g(x_n, y_n, t_n). \qquad (3.2.2b)$$

Multistep methods applied to semi-explicit index one DAE's in this manner are stable and convergent to the same order of accuracy for the DAE as for standard nonstiff ODE's. This result follows simply from the implicit function theorem. Specifically, g can be solved for y_n in terms of x_n and t_n (i.e., $y_n = \tilde{g}(x_n, t_n)$) and then y_n can be inserted into the multistep formula (3.2.2a) to yield the same difference equation as would be obtained by applying the linear multistep method to the ODE

$$x' = f(x, \tilde{g}(x,t), t). \qquad (3.2.3)$$

In principle, a semi-explicit index one DAE can be solved by any linear multistep method which is appropriate for the underlying ODE (3.2.3). In practice, it is important to recognize that if the constraint (3.2.2b) is solved at each step in a separate iteration from the differential equation (3.2.2a), it must be solved with sufficient accuracy so as not to adversely affect the error estimates and other strategies involved in the solution of the differential equation. For this reason, and especially if the underlying ODE (3.2.3) is stiff, it is often advisable to refrain from treating the constraint separately.

3.2.2 Fully-Implicit Index One Systems

In this subsection we develop the main convergence result for BDF methods applied to fully-implicit index one systems.

To begin, we need to define some terminology and prove some preliminary results. Consider the fully-implicit DAE

$$F(t, y, y') = 0, \qquad y(t_0) = y_0. \tag{3.2.4}$$

We assume F is a sufficiently smooth function of its arguments and that there exists a smooth solution $y(t)$ satisfying the given initial values.

Definition 3.2.1 *The nonlinear DAE* (3.2.4) *is said to be* uniform index one *if the index of the constant coefficient system*

$$Aw'(t) + Bw(t) = g(t),$$

where $A = F_{y'}(\hat{t}, \hat{y}, \hat{y}')$, $B = F_y(\hat{t}, \hat{y}, \hat{y}')$, is one for all $(\hat{t}, \hat{y}, \hat{y}')$ in a neighborhood of the graph of the solution, and if

1. *The partial derivatives of A with respect to t, y, y' exist and are bounded in a neighborhood of the solution.*
2. *The rank of A is constant in a neighborhood of the solution.*

We now begin the analysis of the convergence of BDF methods. First we note that without loss of generality, it is sufficient to assume that F has the form

$$F(t, y, y') = \begin{bmatrix} F_1(t, y, y') \\ G(t, y) \end{bmatrix}. \tag{3.2.5}$$

This follows from an idea of Gear [118]. To understand why this would be true, suppose that $\operatorname{rank}(F_{y'}) = r < m$, where m is the dimension of y. Then there exists a nonsingular $r \times r$ submatrix of $F_{y'}$. Suppose the equations have been numbered so that $\operatorname{rank}(\partial F_1/\partial y') = r$, and F_1 is the first r equations in F. Let F_2 be the last $m - r$ equations in F. Suppose the variables are numbered so that

$$\frac{\partial F_1}{\partial y'} = \begin{bmatrix} \dfrac{\partial F_1}{\partial y_1'} & \dfrac{\partial F_1}{\partial y_2'} \end{bmatrix},$$

where $y = [y_1^T, y_2^T]^T$, and $\partial F_1/\partial y_1'$ is nonsingular. Then by the implicit function theorem F_1 can be solved for y_1' in terms of y_1, y_2, y_2', and the expression for y_1' substituted into the last $m - r$ equations to obtain an implicit relationship between y_1 and y_2. The key observation is that y_2' cannot be involved in this relationship, or we would be able to solve F for additional components of y', contrary to the assumption about the rank of $F_{y'}$. Thus F can be written as (3.2.5).

The rearrangement of equations and variables implied by (3.2.5) may not suffice over the entire interval of integration. However, if F is uniform index one, then we can cover the original interval with a finite set of subintervals where F has this form in a neighborhood of the solution. Over any such subinterval, the BDF solution is not altered by rewriting F in the form (3.2.5).

3.2. BDF CONVERGENCE

Thus if BDF converges for (3.2.5) in a subinterval, then it converges for (3.2.4) in that subinterval. Between intersecting subintervals, the initial values needed by BDF can be taken from the previous subinterval. Since there are only a finite number of subintervals, we can without loss of generality for the purposes of BDF convergence assume that F has the form (3.2.5).

Lemma 3.2.1 *For F uniform index one in the form (3.2.5), where y is an m-dimensional vector, there exist nonsingular matrices $P(t, y(t), y'(t))$ and $Q(t, y(t), y'(t))$ such that*

$$PAQ = \begin{bmatrix} I_{m_1} & 0 \\ 0 & 0 \end{bmatrix} \quad (3.2.6a)$$

$$PBQ = \begin{bmatrix} C(t, y(t), y'(t)) & 0 \\ 0 & I_{m_2} \end{bmatrix}, \quad (3.2.6b)$$

where $A(t, y(t), y'(t)) = F_{y'}(t, y(t), y'(t))$, $B(t, y(t), y'(t)) = F_y(t, y(t), y'(t))$, which satisfy

1. $Q(t, y(t), y'(t))$ and $Q^{-1}(t, y(t), y'(t))$ exist and are bounded for all $(t, y(t), y'(t))$ solving (3.2.4),
2. $Q^{-1}(t_1, y(t_1), y'(t_1))Q(t_2, y(t_2), y'(t_2)) = I_m + O(t_2 - t_1)$
3. $C(t_1, y(t_1), y'(t_1)) = C(t_2, y(t_2), y'(t_2)) + O(t_2 - t_1)$,

where $m = m_1 + m_2$.

Proof. For F in the form (3.2.5), we have

$$A = \frac{\partial F}{\partial y'} = \begin{bmatrix} A_{11} & A_{12} \\ 0 & 0 \end{bmatrix},$$

$$B = \frac{\partial F}{\partial y} = \begin{bmatrix} B_{11} & B_{12} \\ B_{21} & B_{22} \end{bmatrix}.$$

It can be verified by substitution that the following P, Q will always bring (A, B) to Kronecker canonical form

$$Q = \begin{bmatrix} Q_{11} & Q_{12} \\ Q_{21} & Q_{22} \end{bmatrix} = \begin{bmatrix} A_{11} & A_{12} \\ B_{21} & B_{22} \end{bmatrix}^{-1}, \quad (3.2.7a)$$

$$P = \begin{bmatrix} I_{m_1} & -(B_{11}Q_{12} + B_{12}Q_{22}) \\ 0 & I_{m_2} \end{bmatrix}. \quad (3.2.7b)$$

The matrix

$$\begin{bmatrix} A_{11} & A_{12} \\ B_{21} & B_{22} \end{bmatrix}$$

is always guaranteed to be nonsingular for index one systems. This follows directly from the definition of index given in Section 2.2. The requirements on the smoothness of Q, Q^{-1} and C follow from the form of P, Q above together with the smoothness assumptions on F. □

Finally, to prove the main result of this subsection we will need the following lemma, which bounds the powers of the stability matrix for BDF applied to uniform index one DAE's.

Lemma 3.2.2 *Let $\mathcal{K}$ be defined by*

$$\mathcal{K} = \begin{bmatrix} \hat{K}_{11} & \hat{K}_{12} \\ 0 & N \end{bmatrix},$$

where $\hat{K}_{11} = K_{11} + O(h)$, $\hat{K}_{12} = O(h)$, N is nilpotent of order k, and

$$K_{11} = \begin{bmatrix} \gamma_1 I_{m_1} & \gamma_2 I_{m_1} & \cdots & \gamma_k I_{m_1} \\ I_{m_1} & & & \\ & \ddots & & \\ & & I_{m_1} & 0 \end{bmatrix},$$

$$N = \begin{bmatrix} 0 & 0 & \cdots & 0 \\ I_{m_2} & & & \\ & \ddots & & \\ & & I_{m_2} & 0 \end{bmatrix},$$

and γ_i are related to the k-step BDF coefficients α_i by $\gamma_i = -\alpha_i/\alpha_0$, $k < 7$. Let $2k \leq i \leq n$. Then

$$\mathcal{K}^i = \begin{bmatrix} O(1) & O(h) \\ 0 & 0 \end{bmatrix},$$

where $O(h^\delta)$, $\delta = 0, 1$, denotes a matrix whose elements are all bounded by a constant (which is independent of n) times h^δ, and $n \leq 1/h$.

Proof. Let $i = qk + r$, where q, r are integers and $r < k$. By assumption, $q \geq 2$. Then
$$\mathcal{K}^i = \mathcal{K}^{kq} \mathcal{K}^r.$$

Now note that since $N^k = 0$,

$$\mathcal{K}^k = \begin{bmatrix} \hat{K}_{11}^k & \hat{\hat{K}}_{12} \\ 0 & 0 \end{bmatrix},$$

where $\hat{\hat{K}}_{12} = O(h)$. Thus,

$$\mathcal{K}^{kq} = \mathcal{K}^k \mathcal{K}^{k(q-2)} \mathcal{K}^k.$$

3.2. BDF CONVERGENCE

By direct multiplication,

$$\mathcal{K}^{k(q-2)} = \begin{bmatrix} \hat{K}_{11}^{k(q-2)} & \hat{K}_{11}^{k(q-3)} \hat{K}_{12} \\ 0 & 0 \end{bmatrix}.$$

Note that $\hat{K}_{11}$ is the stability matrix for BDF methods applied to standard ODE's, so that $\hat{K}_{11}^{k(q-2)}$ is bounded for $k < 7$, and for $n \leq 1/h$. Recalling that $\hat{K}_{12} = O(h)$, and multiplying on the right by $\mathcal{K}^r$, we have the desired result.□

Now we can state the convergence theorem for k-step BDF ($k < 7$) applied to fully-implicit nonlinear index one DAE's. An outline of the proof is as follows. First a recurrence relation which describes the propagation of errors of the system is derived. Then a time-dependent transformation into Kronecker canonical form, which is possible by Lemma 3.2.1, is made. With this change of coordinates, the stability matrix which propagates the errors is $\mathcal{K}$. Then Lemma 3.2.2 is used to bound the powers of $\mathcal{K}$ and the recurrence relations are solved for the errors. Finally, the higher order nonlinear terms are shown to be small. Note that the k-step BDF method requires k starting values, which are assumed in the theorem to be $O(h^k)$ accurate.

Theorem 3.2.1 *Let (3.2.4) be a uniform index one DAE on an interval $I = [t_0, t_0 + T]$. Then the numerical solution of (3.2.4) by the k-step BDF with fixed stepsize h for $k < 7$ converges to $O(h^k)$ if all initial values are correct to $O(h^k)$ accuracy and if the Newton iteration on each step is solved to $O(h^{k+1})$ accuracy.*

Proof. Let y_n satisfy

$$F\left(t_n, y_n, \frac{1}{h} \sum_{i=0}^{k} \alpha_i y_{n-i}\right) = 0,$$

where the α_i are the BDF coefficients and $t_n = nh$. Without loss of generality, assume F has the form (3.2.5). Let $e_n = y_n - y(t_n)$ and $\rho y_n = \sum_{i=0}^{k} \alpha_i y_{n-i}$ so that

$$\frac{\rho y(t_n)}{h} - y'(t_n) = \frac{\tau_n}{h},$$

where $\tau_n = -\frac{1}{k+1} h^{k+1} y^{(k+1)}(t_n) + O(h^{k+2})$ is the local truncation error. Then

$$\begin{aligned}
0 &= F\left(t_n, y_n, \frac{\rho y_n}{h}\right) \\
&= F\left(t_n, y(t_n) + e_n, \frac{\rho y(t_n)}{h} + \frac{\rho e_n}{h}\right) \\
&= F\left(t_n, y(t_n) + e_n, y'(t_n) + \frac{\tau_n}{h} + \frac{\rho e_n}{h}\right).
\end{aligned}$$

Thus,

$$0 = F(t_n, y(t_n), y'(t_n)) + \frac{\partial F}{\partial y}e_n + \frac{\partial F}{\partial y'}\left(\frac{\tau_n}{h}\right) + \frac{\partial F}{\partial y'}\left(\frac{\rho e_n}{h}\right) + \eta_n, \quad (3.2.8)$$

where $\partial F/\partial y$, $\partial F/\partial y'$ are evaluated at $(t_n, y(t_n), y'(t_n))$, and η_n is composed of residuals from the Newton iteration and higher order terms.

Let $A_n = \partial F/\partial y'$, $B_n = \partial F/\partial y$. Then we can rewrite (3.2.8) as

$$A_n \rho e_n + h B_n e_n + A_n \tau_n + h \eta_n = 0. \quad (3.2.9)$$

Let (P_n, Q_n) transform (A_n, B_n) to the Kronecker canonical form (3.2.6). Let $\tilde{e}_n = Q_n^{-1} e_n$, $\tilde{\tau}_n = Q_n^{-1} \tau_n$, $\tilde{\eta}_n = P_n \eta_n$. Define the matrices S_n, T_n by

$$S_n = Q_n^{-1}\left(A_n + \frac{h}{\alpha_0}B_n\right)^{-1} A_n Q_n = \begin{bmatrix} \left(I_{m_1} + \frac{h}{\alpha_0}C_n\right)^{-1} & 0 \\ 0 & 0 \end{bmatrix}$$

$$T_n = Q_n^{-1}\left(A_n + \frac{h}{\alpha_0}B_n\right)^{-1} P_n^{-1} = \begin{bmatrix} \left(I_{m_1} + \frac{h}{\alpha_0}C_n\right)^{-1} & 0 \\ 0 & \frac{\alpha_0}{h}I_{m_2} \end{bmatrix}.$$

Then (3.2.9) can be written

$$\tilde{e}_n = S_n \sum_{i=1}^{k} \gamma_i (Q_n^{-1} Q_{n-i}) \tilde{e}_{n-i} - \left(\frac{1}{\alpha_0}\right) S_n \tilde{\tau}_n - \frac{h}{\alpha_0} T_n \tilde{\eta}_n, \quad (3.2.10)$$

where the γ_i are related to the BDF coefficients α_i by $\gamma_i = -\alpha_i/\alpha_0$. Finally, partition $\tilde{e}_n = \begin{bmatrix} \tilde{e}_n^{(1)} \\ \tilde{e}_n^{(2)} \end{bmatrix}$, where $\tilde{e}_n^{(1)}$ has dimension m_1 and $\tilde{e}_n^{(2)}$ has dimension m_2. Form $\underline{\tilde{e}}_n = [\tilde{e}_n^{(1)}, \tilde{e}_{n-1}^{(1)}, \ldots, \tilde{e}_{n-k+1}^{(1)}, \tilde{e}_n^{(2)}, \tilde{e}_{n-1}^{(2)}, \ldots, \tilde{e}_{n-k+1}^{(2)}]$. Similarly, partition $\tilde{\eta}_n = \begin{bmatrix} \tilde{\eta}_n^{(1)} \\ \tilde{\eta}_n^{(2)} \end{bmatrix}$ and $\tilde{\tau}_n = \begin{bmatrix} \tilde{\tau}_n^{(1)} \\ \tilde{\tau}_n^{(2)} \end{bmatrix}$. Then we can rewrite (3.2.10) to obtain, after some rearrangement,

$$\underline{\tilde{e}}_n = \mathcal{K}\underline{\tilde{e}}_{n-1} - \left(\frac{1}{\alpha_0}\right)\left[\left(I_{m_1} + \frac{h}{\alpha_0}C_n\right)^{-1}\tilde{\tau}_n^{(1)}, 0, 0, \ldots, 0\right]^T$$

$$- \left(\frac{h}{\alpha_0}\right)\left[\left(I_{m_1} + \frac{h}{\alpha_0}C_n\right)^{-1}\tilde{\eta}_n^{(1)}, 0, 0, \ldots, 0,\right.$$

$$\left.\left(\frac{\alpha_0}{h}\right)\tilde{\eta}_n^{(2)}, 0, 0, \ldots, 0\right]^T. \quad (3.2.11)$$

The error propagation matrix $\mathcal{K}$ is given by

$$\mathcal{K} = \begin{bmatrix} \hat{K}_{11} & \hat{K}_{12} \\ 0 & N \end{bmatrix},$$

3.2. BDF CONVERGENCE

where $\hat{K}_{11} = K_{11} + O(h)$, $\hat{K}_{12} = O(h)$, K_{11} is the BDF error propagation matrix, and N is a nilpotent matrix of order k as described in Lemma 3.2.2. Solving the recurrence (3.2.11) we obtain

$$\begin{aligned}\tilde{\underline{e}}_n &= \mathcal{K}^n \tilde{\underline{e}}_0 - \frac{1}{\alpha_0} \sum_{i=0}^{n-1} \mathcal{K}^i \left[\left(I_{m_1} + \frac{h}{\alpha_0} C_{n-i} \right)^{-1} \tilde{r}_{n-i}^{(1)}, 0, 0, \ldots, 0 \right]^T \\ &\quad - \frac{h}{\alpha_0} \sum_{i=0}^{n-1} \mathcal{K}^i \left[\left(I_{m_1} + \frac{h}{\alpha_0} C_{n-i} \right)^{-1} \tilde{\eta}_{n-i}^{(1)}, 0, 0, \ldots, 0, \right. \\ &\quad \left. \left(\frac{\alpha_0}{h} \right) \tilde{\eta}_{n-i}^{(2)}, 0, 0, \ldots, 0 \right]^T. \end{aligned} \quad (3.2.12)$$

Using Lemma 3.2.2 to bound the size of the blocks of $\mathcal{K}^i$ for $2k \leq i \leq n$, and noting the size of the blocks of $\mathcal{K}^i$ for $i < 2k$, we have from (3.2.12)

$$\left\| \tilde{e}_n^{(1)} \right\| = O(\|\tilde{\underline{e}}_0\|) + O(h^k) + O(\|\tilde{\eta}_n^{(1)}\|) + O(\|\tilde{\eta}_n^{(2)}\|) \quad (3.2.13a)$$

$$\left\| \tilde{e}_n^{(2)} \right\| = O(\|\tilde{\eta}_n^{(2)}\|). \quad (3.2.13b)$$

For linear problems, $\tilde{\eta}$ is composed of residuals from the Newton iteration which are $O(h^{k+1})$, and we are done. For nonlinear problems, we must deal with the higher order terms $\tilde{\eta}$. Noting the form of $\mathcal{K}$, the recurrence (3.2.11) gives an expression for $\|\rho \tilde{e}_n^{(1)}/h\|$. Dividing (3.2.13b) by h yields a bound on the size of $\|\rho \tilde{e}_n^{(2)}/h\|$, which because of errors in the initial values is valid for $n \geq k+1$.

$$\begin{aligned}\left\| \frac{\rho \tilde{e}_n^{(1)}}{h} \right\| &= O(\|\tilde{\underline{e}}_0\|) + O(h^k) + O(\|\tilde{\eta}_n^{(1)}\|) + O(\|\tilde{\eta}_n^{(2)}\|) \\ \left\| \frac{\rho \tilde{e}_n^{(2)}}{h} \right\| &= O\left(\frac{\|\tilde{\eta}_n^{(2)}\|}{h} \right). \end{aligned} \quad (3.2.14)$$

Since F has the form (3.2.5), $\eta_n^{(1)}$ consists of residuals from the Newton iteration which are $O(h^{k+1})$ and higher order terms of the form

$$e_n^T F_{yy} e_n, \quad \left(\frac{\rho e_n^T}{h} \right) F_{y'y} e_n, \quad \left(\frac{\rho e_n^T}{h} \right) F_{y'y'} \left(\frac{\rho e_n}{h} \right),$$

and $\eta_n^{(2)}$ consists of residuals from the Newton iteration and terms of the form

$$e_n^T F_{yy} e_n.$$

Because P_n has the form given by (3.2.7b), and $\tilde{\eta}_n = P_n \eta_n$, it follows that $\tilde{\eta}_n^{(1)}$ and $\tilde{\eta}_n^{(2)}$ are composed of terms of the same form as $\eta_n^{(1)}$ and $\eta_n^{(2)}$, respectively.

Suppose that the nonlinear higher order terms satisfy $\|\tilde{\eta}_n^{(i)}\| \leq \epsilon_i$, $i = 1, 2$ (this does not include the Newton errors, which are assumed to be $O(h^{k+1})$), and the errors in the initial conditions satisfy $\|\tilde{e}_0\| = O(h^k)$. Then we will show that ϵ_i are small, to conclude the proof. According to the form of $\tilde{\eta}_n^{(1)}$ and $\tilde{\eta}_n^{(2)}$, and substituting from (3.2.13) and (3.2.14), for $n \geq k+1$ we have

$$\|\tilde{\eta}_n^{(1)}\| \leq K_1 \left(h^k + \epsilon_1 + \epsilon_2 + \frac{\epsilon_2}{h} \right)^2$$
$$\|\tilde{\eta}_n^{(2)}\| \leq K_2 \left(h^k + \epsilon_1 + \epsilon_2 \right)^2.$$

Now, determine ϵ_1 and ϵ_2 as solutions of

$$\epsilon_1 = K_3 \left(h^{2k} + \epsilon_1^2 + \epsilon_2^2 + \frac{\epsilon_1 \epsilon_2}{h} + \frac{\epsilon_2^2}{h} + \frac{\epsilon_2^2}{h^2} + h^k \epsilon_1 + h^k \epsilon_2 + h^{k-1} \epsilon_2 \right)$$
$$\epsilon_2 = K_4 \left(h^{2k} + \epsilon_1^2 + \epsilon_2^2 + \epsilon_1 \epsilon_2 + h^k \epsilon_1 + h^k \epsilon_2 \right), \qquad (3.2.15)$$

by solving (3.2.15) by functional iteration $\underline{\epsilon} = G(\underline{\epsilon})$ with initial value $\underline{\epsilon}^{(0)}$ satisfying

$$\underline{\epsilon}^{(0)} = O(h^{2k}).$$

For $k \geq 2$, we can use the contraction mapping theorem to conclude $\underline{\epsilon} = O(h^{2k})$, and we are done. For $k = 1$, we cannot apply the theorem directly, because $\|\partial G/\partial \underline{\epsilon}\| = O(1)$. But if we scale the variables by $\bar{\epsilon}_1 = \epsilon_1$, $\bar{\epsilon}_2 = \epsilon_2/\sqrt{h}$, then we can apply the strategy above. Finally, the result now follows from (3.2.13). □

It can be shown, using an argument quite similar to the one we have given here, that if *variable stepsize* BDF methods are implemented in such a way that the method is stable for standard ODE's, then the k-step BDF method, ($k < 7$) is convergent for fully-implicit index one DAE's. This result for variable stepsize BDF applied to fully-implicit index one systems also follows directly from a result for semi-explicit index two DAE's which was originally given in [121]. In [47,82,124,125,133], conditions on the sequences of stepsizes for various implementations of variable stepsize BDF which guarantee stability as the maximum stepsize tends to zero are given.

The convergence and order results for constant stepsize and variable stepsize BDF are important for practical applications because together they justify the methods and some of the strategies employed in BDF DAE codes such as LSODI [143] and DASSL [201].

3.2.3 Semi-Explicit Index Two Systems

In this subsection we study the behavior of BDF methods applied to semi-explicit index two systems. The main result is a convergence theorem which shows that the k-step BDF method ($k < 7$) is globally convergent to $O(h^k)$. Historically, this result was obtained independently of the index one results [30,33,170]. Using an observation of Gear [118], we are able to give a very

3.2. BDF CONVERGENCE

simple proof which follows from the index one BDF convergence analysis of the previous subsection.

Consider the semi-explicit nonlinear system

$$f(x, x', y, t) = 0 \qquad (3.2.16a)$$
$$g(x, y, t) = 0, \qquad (3.2.16b)$$

where we assume the system has index two, $(\partial f/\partial x')^{-1}$ exists and is bounded in a neighborhood of the solution, $\partial g/\partial y$ has constant rank, and f and g have as many continuous partial derivatives as desired in a neighborhood of the solution.

Theorem 3.2.2 *Suppose the nonlinear semi-explicit index two system (3.2.16), is to be solved numerically by the k-step BDF method ($k < 7$), the errors in the initial values are $\|e_0\| = O(h^k)$, and the errors in terminating the Newton iteration satisfy $O(h^{k+1})$. Then the k-step BDF method is convergent and globally accurate to $O(h^k)$, after $k+1$ steps.*

Proof. For the semi-explicit nonlinear index two system (3.2.16), let $w' = y$ and consider the related nonlinear index one system

$$f(x, x', w', t) = 0 \qquad (3.2.17a)$$
$$g(x, w', t) = 0. \qquad (3.2.17b)$$

It is easy to see that solving the index one system (3.2.17) by the k-step BDF method gives exactly the same solution for x as solving the original semi-explicit index two system (3.2.16) by the k-step BDF method. Thus for linear problems the solution for x is accurate to $O(h^k)$.

Now consider the error in y. The BDF discretization of (3.2.17) is given by

$$f\left(x_n, \frac{\rho x_n}{h}, \frac{\rho w_n}{h}, t_n\right) = 0 \qquad (3.2.18a)$$
$$g\left(x_n, \frac{\rho w_n}{h}, t_n\right) = 0. \qquad (3.2.18b)$$

Let $e_n^w = w_n - w(t_n)$. Then by (3.2.14) we have, for $n \geq k+1$,

$$\left\|\frac{\rho e_n^w}{h}\right\| = O(h^k). \qquad (3.2.19)$$

Now, what is really of interest is the error in y. Let $e_n^y = y_n - y(t_n)$. Then we have

$$y_n = \frac{\rho w_n}{h} = \frac{\rho w(t_n)}{h} + \frac{\rho e_n^w}{h}. \qquad (3.2.20)$$

The true solution satisfies

$$y(t_n) = w'(t_n) \qquad (3.2.21)$$

Subtracting (3.2.21) from (3.2.20) and noting that $w'(t_n) - \rho w(t_n)/h$ is $O(h^k)$ for the k-step BDF method, we obtain

$$e_n^y = \frac{\rho e_n^w}{h} + O(h^k) = O(h^k). \quad \square$$

We note that for starting values which are accurate to $O(h^{k+1})$, the $O(h^k)$ convergence takes place immediately, rather than after $k+1$ steps [33]. This is likely to be the case in practice, where the starting values are usually computed much more accurately than the subsequent solution, to guard against failures of the stepsize control strategies in an automatic code.

It can easily be shown in a similar way that if variable stepsize BDF methods are implemented in such a way that the method is stable for standard ODE's, then the k-step BDF method ($k < 7$) is convergent for semi-explicit index two DAE's. Originally, this result was given by Gear, Gupta and Leimkuhler in [121]. Conditions on the implementation which guarantee stability as the maximum stepsize tends to zero are discussed in [47,82,124,125,133].

3.2.4 Index Three Systems of Hessenberg Form

One of the interesting properties of BDF methods is that while they were not originally formulated with the solution of DAE systems in mind, they suffer no loss of accuracy for many DAE systems, including higher index systems. Earlier in this chapter we have seen that the k-step BDF solution is accurate to $O(h^k)$ for nonlinear index one systems, semi-explicit nonlinear index two systems, and for linear constant coefficient systems of any index (after an initial boundary layer). In this subsection we will discuss some convergence results for BDF methods applied to nonlinear index three systems of Hessenberg form. These are systems which can be written as

$$\begin{aligned} x_1' &= F_1(x_1, x_2, x_3, t) \\ x_2' &= F_2(x_1, x_2, t) \\ 0 &= F_3(x_2, t), \end{aligned} \quad (3.2.22)$$

where the system is index three if the matrix $(\partial F_3/\partial x_2)(\partial F_2/\partial x_1)(\partial F_1/\partial x_3)$ is nonsingular. Systems of this form arise typically in the modeling of constrained mechanical systems and of trajectory prescribed path control problems. We discuss these applications in some detail in Chapter 6.

Convergence of BDF methods for (3.2.22) has been studied in [30,33,170]. The most general result is given by Brenan and Engquist [33]. We state this result here. First, we need to define some terminology. A set of k starting values $(x_1^{(j)}, x_2^{(j)}, x_3^{(j)})$ at $t_0 + jh$, $j = 0, 1, \ldots, k-1$ is said to be *numerically consistent* to order $k+1$ if there exists a solution to (3.2.22) such that

$$\begin{aligned} \|x_1^{(j)} - x_1(t_j)\| &= O(h^{k+1}) \\ \|x_2^{(j)} - x_2(t_j)\| &= O(h^{k+1}) \\ \|F_3\left(x_2^{(j)}, t_j\right)\| &= O(h^{k+2}). \end{aligned} \quad (3.2.23)$$

3.2. BDF CONVERGENCE

The starting values for x_3 are not critical in the convergence analysis for (3.2.22).

Theorem 3.2.3 *There exists a numerical solution of the index three system (3.2.22) by the k-step BDF with constant stepsize h for $k < 7$ which converges globally with kth order accuracy to a solution of (3.2.22) after $k + 1$ steps if the starting values are numerically consistent to order $k + 1$, and the algebraic equations at each step are solved to $O(h^{k+2})$ accuracy for $k \geq 2$, and $O(h^{k+3})$ accuracy for $k = 1$.*

In contrast to the situation for semi-explicit index two systems, the $O(h^k)$ convergence results for the k-step BDF method applied to index three systems cannot be extended to hold for variable stepsize meshes because each time the stepsize is changed, a new boundary layer of reduced convergence rates is initiated. In particular, the first order BDF (implicit Euler method) fails to converge at the end of the first step following a change in the stepsize.

To gain a better understanding of why there would be an order reduction for variable stepsizes, consider the implicit Euler method applied to the canonical index three system

$$\begin{aligned} x'_1 &= x_2 \\ x'_2 &= x_3 \\ 0 &= x_1 - g(t). \end{aligned}$$

After three steps, we have

$$\begin{aligned} x_{1,n+1} &= g(t_{n+1}) \\ x_{2,n+1} &= \frac{1}{h_{n+1}} \left(g(t_{n+1}) - g(t_n) \right) \\ x_{3,n+1} &= \frac{1}{h_{n+1}} \left(\left(\frac{g(t_{n+1}) - g(t_n)}{h_{n+1}} \right) - \left(\frac{g(t_n) - g(t_{n-1})}{h_n} \right) \right). \end{aligned}$$

Now, while $x_{3,n+1}$ converges to $g''(t_{n+1})$ to $O(h^2)$ if the stepsize is constant, it *blows up* as $h_{n+1} \to 0$ for h_n fixed!

The situation with respect to convergence for higher order BDF with variable stepsizes is not quite so discouraging as for implicit Euler. At the time of this writing, extensive numerical experiments [120] have demonstrated that the k-step variable stepsize BDF gives order $k - \nu + 2$ for index ν DAE's in Hessenberg form, provided the stepsize sequence is stable for ODE's. This order behavior has been shown for constant coefficient DAE systems [119], and we conjecture that it is true for general Hessenberg systems. Still, at the present time we feel that we cannot recommend solving general nonlinear Hessenberg systems directly by variable stepsize BDF codes because of difficulties with the implementation in an automatic code and extreme sensitivity of the higher order BDF solutions to the accuracy of the starting values (see Section

6.3 for a numerical example). BDF solutions to some special classes of problems such as constrained mechanical systems, where the Lagrange multipliers appear linearly in the system, seem to be somewhat more robust but are still not completely reliable.

We conclude this section with two other observations about BDF methods. First, for higher index linear time invariant systems $Ax' + Bx = f$, the inconsistent initial conditions correspond to discontinuous and distributional solutions [49]. Earlier in this chapter we showed that BDF methods applied to higher index DAE's have an initial boundary layer of nonconvergence (with respect to the sup norm). However, the implicit Euler method applied to $Ax' + Bx = f$ actually converges for all initial conditions on the entire time interval independent of the index [61]. For consistent initial conditions, the error in the boundary layer is $O(h)\delta(h)$ where $\delta(h)$ is a distribution that varies continuously in h. Thus the error goes to zero weakly. For inconsistent initial conditions the error has the same form, but now the one-step BDF approximation converges to a distributional solution. The proof is such that a similar result will hold for the k-step BDF method. While primarily of theoretical interest, these results help in understanding our previous results on BDF methods.

Secondly, we have developed the BDF theory in this chapter for solvable systems which are the kind usually encountered in practice. It is interesting to consider what would happen if a BDF method was applied to a DAE which was not solvable but for which the method was implementable. An example is given in [64] of a linear time varying DAE, $A(t)x' + B(t)x = f$, which is not solvable in that solutions are not uniquely determined by consistent initial conditions. However, the BDF methods are implementable since the pencil (A, B) is regular for every t. For this example, it is shown that for the same initial condition, and any choice of initial values $x(t_0)$, $x(t_0 + h)$ for the two-step method, that the one-step BDF and two-step BDF approximations will then converge to different solutions.

3.3 BDF Methods, DAE's, and Stiff Problems

We have already commented that DAE's have some characteristics in common with stiff differential equations. One source of DAE's is the reduced order problem in a singular perturbation analysis of a stiff differential equation. These DAE's are important both as models in their own right, and for giving insight about the original stiff model. Similarly, examining the relationship between the numerical solution of the DAE and the numerical solution of the stiff problem provides insight into the numerical solution of both types of problems. The relationship between implicit Runge-Kutta methods applied to stiff ODE's and their application to DAE's will be briefly discussed in Section 4.5. In this section we shall consider BDF methods. It will be shown that just as the analytic solution can be written as an asymptotic expansion with

3.3. BDF METHODS, DAE'S, AND STIFF PROBLEMS

the analytic solution of the DAE the first order term of the outer part of the expansion, it is also possible to write the numerical solution of the stiff problem as an asymptotic expansion with the numerical solution of the DAE as the first-order term in the outer part of the expansion. The discussion and results will be based on the work of Lötstedt [167,168].

Index One Systems

We shall first review the usual singular perturbation theory in order to motivate the development of the numerical results. The full, or regular, problem will taken to be in the form

$$\begin{aligned} x' &= f(x,y,t,\epsilon), & x(0) &= \xi_0 \\ \epsilon y' &= g(x,y,t,\epsilon), & y(0) &= \eta_0 \end{aligned} \quad (3.3.1)$$

on the interval $\mathcal{I} = [0,T]$ with ϵ a small positive parameter. The reduced problem is then the semi-explicit DAE

$$\begin{aligned} \overline{x}' &= f(\overline{x},\overline{y},t,0), & \overline{x}(0) &= \xi_0 \\ 0 &= g(\overline{x},\overline{y},t,0). \end{aligned} \quad (3.3.2)$$

We assume that (3.3.2) has a solution on $\mathcal{I}$, that f, g have continuous second partials in a neighborhood of this solution, and that $\partial g/\partial y$ is uniformly negative definite on this neighborhood. Then (3.3.2) is index one and the solution of (3.3.1) converges to that of (3.3.2) as $\epsilon \to 0^+$. To be more precise, let $x(t,\epsilon)$, $y(t,\epsilon)$ be the solution of (3.3.1). Then

$$\begin{aligned} x(t,\epsilon) &= x^\epsilon(t) + X^\epsilon\left(\tfrac{t}{\epsilon}\right) \\ y(t,\epsilon) &= y^\epsilon(t) + Y^\epsilon\left(\tfrac{t}{\epsilon}\right) \end{aligned} \quad (3.3.3)$$

where

$$\begin{aligned} x^\epsilon(t) &= \overline{x}(t) + \epsilon\tilde{x}(t,\epsilon) \\ y^\epsilon(t) &= \overline{y}(t) + \epsilon\tilde{y}(t,\epsilon) \\ X^\epsilon(\tau) &= \overline{X}(\tau) + \epsilon\tilde{X}(\tau,\epsilon) \\ Y^\epsilon(\tau) &= \overline{Y}(\tau) + \epsilon\tilde{Y}(\tau,\epsilon). \end{aligned}$$

The terms with a tilde are bounded uniformly in ϵ, t for $0 \leq t \leq T$, $0 \leq \epsilon \leq \epsilon_0$ for some ϵ_0. The vector (x^ϵ, y^ϵ) is sometimes called the *outer solution* and describes the solution away from $t = 0$, while (X^ϵ, Y^ϵ) is called the *inner solution* and gives transient terms in an initial boundary layer. For (3.3.1) we have that $\overline{X}(\tau), \overline{Y}(\tau), 0 \leq \tau \leq t/\epsilon$ are given by the differential equation

$$\begin{aligned} \frac{d\overline{X}}{d\tau} &= 0 \\ \frac{d\overline{Y}}{d\tau} &= g(\overline{X},\overline{Y},0,0). \end{aligned}$$

By the assumption on $\partial g/\partial y$ we know that $\lim_{\tau\to\infty} \overline{Y}(\tau) = 0$ so that $\overline{Y}(t/\epsilon)$ becomes small as t increases, or ϵ decreases. More precisely, there are constants c, δ, ϵ' so that if $0 < \epsilon < \epsilon'$, then

$$\|X^\epsilon(t/\epsilon) + Y^\epsilon(t/\epsilon)\| \le c\|y(0) - y^\epsilon(0)\|e^{-\delta t/\epsilon}. \tag{3.3.4}$$

Now let h be a fixed stepsize, $t_n = nh$. Let $\rho x_n = \sum_{i=0}^{k} \alpha_i x_{n-i}$ where $\rho x_n/h$ is the BDF operator. Applying a k-step BDF method to (3.3.1) and (3.3.2) results in

$$\begin{aligned}\rho x_n &= hf(x_n, y_n, t_n, \epsilon), & x_0 &= \xi_0 \\ \epsilon\rho y_n &= hg(x_n, y_n, t_n, \epsilon), & y_0 &= \eta_0\end{aligned} \tag{3.3.5}$$

and

$$\begin{aligned}\rho \overline{x}_n &= hf(\overline{x}_n, \overline{y}_n, t_n, \epsilon), & \overline{x}_0 &= \xi_0 \\ 0 &= hg(\overline{x}_n, \overline{y}_n, t_n, 0),\end{aligned} \tag{3.3.6}$$

respectively.

In [167,168], the first q values, $k - 1 \le q \le n$, are found using an implicit Euler method. Then a k-step BDF method is used for the remainder of the interval. In general, of course, this would be expected to result in a first-order method globally. Since our intent is to emphasize the flavor of the results, we shall take $k = 1$ on the entire interval $\mathcal{I}$.

We may write the solutions of the BDF equations (3.3.5) and (3.3.6) as

$$\begin{aligned}x_n &= x_n^\epsilon + X_n^\epsilon \\ y_n &= y_n^\epsilon + Y_n^\epsilon\end{aligned}$$

where

$$\begin{aligned}x_n^\epsilon &= \overline{x}_n + \epsilon\tilde{x}(n,\epsilon) \\ y_n^\epsilon &= \overline{y}_n + \epsilon\tilde{y}(n,\epsilon) \\ X_n^\epsilon &= \overline{X}_n + \epsilon\tilde{X}(n,\epsilon) \\ Y_n^\epsilon &= \overline{Y}_n + \epsilon\tilde{Y}(n,\epsilon).\end{aligned}$$

Letting $H = h/\epsilon$, $\tau_n = nH$, we have that $\overline{X}_n, \overline{Y}_n$ are given by the difference equation

$$\begin{aligned}\rho\overline{X}_n &= 0 \\ \rho\overline{Y}_n &= Hg(\overline{x}_n, \overline{y}_n, \overline{X}_n, \overline{Y}_n, t_n, 0).\end{aligned}$$

One interesting aspect of this study concerns the interplay between ϵ and h.

Theorem 3.3.1 *Suppose that f, g satisfy the assumptions made earlier in this section. Then for a given $h > 0$, there is an $\tilde{\epsilon} > 0$ such that if $0 \le \epsilon \le \tilde{\epsilon} \ll h$, then the expansions (3.3.3) are uniformly valid. That is, the tilde terms are uniformly bounded. Furthermore there are constants c, Θ with $\Theta < 1$ such that*

$$\|\overline{Y}_n\| \le \|\overline{y}_0 - y(0)\|c\Theta^n. \tag{3.3.7}$$

3.3. BDF METHODS, DAE'S, AND STIFF PROBLEMS

Note that (3.3.7) is the discrete analogue of (3.3.4). This theorem establishes a rigorous relationship between the one-step BDF solution of the stiff system (3.3.1) and the one-step BDF solution of the DAE that results when $\epsilon = 0$.

Index Two Systems

Lötstedt also considers the index two DAE
$$\begin{aligned}\overline{x}' &= f(\overline{x},\overline{y},t,0), \quad \overline{x}(0) = \xi_0 \\ 0 &= g(\overline{x},t,0)\end{aligned} \qquad (3.3.8)$$

with $(\partial g/\partial x)(\partial f/\partial y)$ nonsingular, which is a Hessenberg system of order two. If one wishes to view (3.3.8) as a reduced order equation for a perturbed problem, and if the solutions are to be continuous in the perturbation parameter, then the obvious regularization

$$\begin{aligned}x' &= f(x,y,t,\epsilon) \\ \epsilon y' &= g(x,t,\epsilon)\end{aligned}$$

may not be the correct regularization. This is the case even for the linear time invariant system
$$Ax' + Bx = f(t). \qquad (3.3.9)$$

Cobb [78,79] has shown that a correct regularization for (3.3.9) is the *pencil regularization*
$$(A + \epsilon B)x' + Bx = f(t), \qquad (3.3.10)$$
in that the solutions of (3.3.10) converge to those of (3.3.9) as ϵ goes to zero. The regularization (3.3.10) has more than two time scales if the index of (3.3.9) is greater than one. Rather than attempt a full regularization of (3.3.8), with its multiple time scales, Lötstedt considers a partial regularization. As we shall see shortly, this approach is closely related to certain other ideas in the numerical literature.

We begin by introducing the system
$$\begin{aligned}x' &= f(x,y,t,\epsilon), \quad x(0) = \xi_0 & (3.3.11\text{a}) \\ \epsilon y &= g(x,t,\epsilon), \quad y(0) = \frac{g(\xi_0,0,\epsilon)}{\epsilon}. & (3.3.11\text{b})\end{aligned}$$

This system is a *singular singular perturbation problem* [196] since for $\epsilon > 0$, it is an index one DAE, while for $\epsilon = 0$ it is index two.

Differentiating (3.3.11b) and using (3.3.11a) we get the ODE
$$\begin{aligned}x' &= f(x,y,t,\epsilon), \quad x(0) = \xi_0 & (3.3.12\text{a})\\ \epsilon y' &= g_x(x,t,\epsilon)f(x,y,t,\epsilon) + g_t(x,t,\epsilon), \quad y(0) = \frac{g(\xi_0,0,\epsilon)}{\epsilon} & (3.3.12\text{b})\end{aligned}$$

and the corresponding reduced order problem

$$\bar{x}' = f(\bar{x},\bar{y},t,0), \quad x(0) = \xi_0 \quad (3.3.13a)$$
$$0 = g_x(\bar{x},t,0)f(\bar{x},\bar{y},t,0) + g_t(\bar{x},t,0). \quad (3.3.13b)$$

This pair of systems is not the same as the pair examined in the last section because of the initial condition $y(0) = g(\xi_0, 0, \epsilon)/\epsilon$ in (3.3.12b). This initial condition is used so that if the initial condition $x(0) = \xi_0$ is consistent for the index two DAE (3.3.8), it will also be consistent for the index one regularization (3.3.11).

Note also that the solutions of the original index two DAE (3.3.8) are solutions of the index one DAE (3.3.13), and that the solutions of (3.3.11) are solutions of (3.3.12).

Assuming that $(\partial g/\partial x)(\partial f/\partial y)$ is negative definite in a neighborhood of the solution of the initial problem, an analysis can be carried out for these systems that is similar to that for index one systems. If $\|g(\xi_0, 0, \epsilon)\| = O(\epsilon)$, then uniform asymptotic expansions exist. If, however, $\|g(\xi_0, 0, \epsilon)\| = O(1)$, which would be the case if $g(\xi_0, 0, 0) \neq 0$, then there is a term which has a distributional limit as $\epsilon \to 0^+$. This is an index two nonlinear semi-explicit version of the general linear results of Cobb.

For our later analysis, we need to note that, by the implicit function theorem, we can solve (3.3.13b) locally to get $\bar{y} = \psi(t, \bar{x})$ so that (3.3.13a) becomes

$$\bar{x}' = f(\bar{x}, \psi(t, \bar{x}), t, 0).$$

We need the additional property that there is a suitable Liapunov function for

$$z' = f(\bar{x} + z, \psi(\bar{x} + z), t), t, 0) - f(\bar{x}, \psi(\bar{x}, t), t, 0),$$

which is the differential equation for $x - \bar{x}$.

With this understanding of the continuous problem, we turn to the corresponding BDF discretizations

$$\rho \bar{x}_n = hf(\bar{x}_n, \bar{y}_n, t_n, 0), \quad \bar{x}_0 = \xi_0$$
$$0 = g(\bar{x}_n, t_n, 0)$$

and

$$\rho x_n = hf(x_n, y_n, t_n, \epsilon), \quad x_0 = \xi_0$$
$$\epsilon y_n = g(x_n, t_n, \epsilon), \quad y_0 = \frac{g(\epsilon, 0, \epsilon)}{\epsilon}.$$

Under the previous analytic assumptions, we have that $\|x_n - \bar{x}_n\| = O(h)$ after a fixed number of steps and x_n satisfies an estimate of the form

$$\|g(x_n, t_n, \epsilon)\| < \epsilon c \left(1 + \frac{\eta}{h}\right).$$

3.4. GENERAL LINEAR MULTISTEP METHODS

To illustrate how these results could be useful, suppose that we have the ODE
$$x' = f(x) \tag{3.3.14}$$
whose solutions satisfy some constraint $g(x) = 0$, such as conservation of energy. The numerical solution of (3.3.14) need not satisfy the constraint. Several different approaches have been formulated to circumvent this problem. One, due to Baumgarte [14], is to set up the DAE
$$\begin{aligned} x' &= f(x) + g_x^T \mu \\ \epsilon \mu &= g(x) \end{aligned} \tag{3.3.15}$$
with $\epsilon > 0$. The only solutions of (3.3.15) are x a solution of (3.3.14) and $\mu = 0$.

Treating μ as an unknown function, we find that
$$\mu' = -\frac{1}{\epsilon}(g_x g_x^T)\mu + \frac{1}{\epsilon} g_x f.$$

If $g_x g_x^T$ is positive definite and $g_x f$ satisfies suitable assumptions, then we have that $\mu \to 0$. Notice that (3.3.15) is in the form of (3.3.11) with $y = \mu$. In this context, Lötstedt's results say that by taking h small we can compute x to $O(h)$ accuracy with the one-step BDF method. If we then take $\hat{\epsilon} = \epsilon/h$ small we can also satisfy the constraint to $O(\epsilon/h)$ accuracy. The impulsive behavior, and hence the boundary layer, is in the μ variable which is not of major concern in this problem.

3.4 General Linear Multistep Methods

In this section, we outline the work of Griepentrog and März [132] on the application of general linear multistep and one-leg methods to index one DAE systems.

For the nonlinear ODE
$$x' = g(x, t), \tag{3.4.1}$$
a *linear multistep* method takes the form
$$\frac{1}{h} \sum_{j=0}^{k} a_j x_{n-j} - \sum_{j=0}^{k} b_j g(x_{n-j}, t_{n-j}) = 0. \tag{3.4.2}$$

A *one-leg* method for (3.4.1) involves only one evaluation of g and takes the form
$$\frac{1}{h} \sum_{j=0}^{k} a_j x_{n-j} - g\left(\sum_{j=0}^{k} b_j x_{n-j}, \sum_{j=0}^{k} b_j t_{n-j}\right) = 0. \tag{3.4.3}$$

Note that a BDF method can be considered as both a one-leg method and as a multistep method by taking $b_0 = 1$ and the rest of the $b_i = 0$.

For one-leg methods, there is a natural extension of (3.4.3) to the fully-implicit DAE
$$F(t, x, x') = 0. \qquad (3.4.4)$$
The extension is
$$F\left(\sum_{j=0}^{k} b_j t_{n-j}, \sum_{j=0}^{k} b_j x_{n-j}, \frac{1}{h}\sum_{j=0}^{k} a_j x_{n-j}\right) = 0. \qquad (3.4.5)$$

The multistep methods (3.4.2) do not have an immediate extension to the fully-implicit DAE (3.4.4) since these methods require a linear combination of several evaluations of a formula for x' in terms of x, t and that formula is not immediately available. The situation is a little simpler for the semi-explicit DAE
$$\begin{aligned} x' &= f(x,y,t) \\ 0 &= g(x,y,t). \end{aligned} \qquad (3.4.6)$$
For (3.4.6), two definitions of a multistep method suggest themselves. They are
$$\frac{1}{h}\sum_{j=0}^{k} a_j x_{n-j} - \sum_{j=0}^{k} b_j f(x_{n-j}, y_{n-j}, t_{n-j}) = 0 \qquad (3.4.7a)$$
$$\sum_{j=0}^{k} b_j g(x_{n-j}, y_{n-j}, t_{n-j}) = 0, \qquad (3.4.7b)$$
and, as in Subsection 3.2.1,
$$\frac{1}{h}\sum_{j=0}^{k} a_j x_{n-j} - \sum_{j=0}^{k} b_j f(x_{n-j}, y_{n-j}, t_{n-j}) = 0 \qquad (3.4.8a)$$
$$g(x_n, y_n, t_n) = 0. \qquad (3.4.8b)$$

Clearly (3.4.8) implies (3.4.7). If the starting values $\{(x_i, y_i, t_i) : 0 \leq i \leq k-1\}$ satisfy (3.4.8b) and $b_0 \neq 0$, then (3.4.7) implies (3.4.8). We shall consider (3.4.8).

In order to define a multistep method such as (3.4.8) for the general DAE (3.4.4), it is necessary to have some type of semi-explicit form available which distinguishes algebraic constraints (3.4.8b) from differential equations (3.4.8a), or equivalently, algebraic variables y from differentiated variables x. One option would be to use the ideas of Chapter 2 and rewrite (3.4.4) as a semi-explicit system of one higher index
$$\begin{aligned} x' &= y \\ 0 &= F(t, x, y). \end{aligned} \qquad (3.4.9)$$

However, if (3.4.4) is index one, the new system (3.4.9) is index two, which causes difficulty with the analysis. An alternative is explored in [132] under

3.4. GENERAL LINEAR MULTISTEP METHODS

the assumption that the null space of $F_{x'}$ depends only on t and that $F_{x'}$ has constant rank. The remainder of this subsection will discuss the results of [132] for general one-leg and multistep methods.

Let $Q(t)$ be a projection onto the nullspace of $F_{x'}$, $\mathcal{N}(F_{x'})$, and let $P(t) = I - Q(t)$. The projections P, Q will be as smooth as functions of t as $F_{x'}$ is. The approach of [132] differs slightly from the one that we have taken in the rest of this chapter in that the initial conditions are taken to be

$$x(t_0) - x^0 \in \mathcal{N}(F_{x'}) \tag{3.4.10}$$

or equivalently

$$P(t_0)(x(t_0) - x^0) = 0,$$

whereas the initial conditions in the rest of this chapter are given by

$$x(t_0) = x^0.$$

The formulation (3.4.10) has the advantage that for index one problems there is a unique solution for every x^0. It has the disadvantage that only $P(t_0)x(t_0)$ is given by the initial condition and hence $Q(t_0)x(t_0)$ must be estimated before numerical integration can begin.

The assumptions of [132] are detailed in the next definition. A slight relaxation of some of the assumptions is possible.

Definition 3.4.1 *Let $\mathcal{G}$ be an open connected set in $\mathcal{R}^{2m+1}$. Let $\mathcal{G}_t = \{t \in \mathcal{R} : (t, z) \in \mathcal{G} \text{ for some } z\}$. The DAE (3.4.4) is transferable on $[t_0, T] \subset \mathcal{G}_t$ if the following hold:*

1. *$F_t, F_x, F_{x'}$ are continuous on $\mathcal{G}$,*

2. *The nullspace of $F_{x'}$ does not depend on x, x',*

3. *$F_{x'}$ has constant rank,*

4. *The projector $Q(t)$ onto the nullspace of $F_{x'}$ is smooth,*

5. *$G(t, x, x') = F_{x'} + F_x Q$ is invertible on $([t_0, T] \times \mathcal{R}^{2m}) \cap \mathcal{G}$,*

6. *G^{-1} is bounded on $([t_0, T] \times \mathcal{R}^{2m}) \cap \mathcal{G}$,*

7. *$F(t, x, x') = 0$ has a smooth solution on $[t_0, T]$.*

If extra smoothness is assumed on F in (1) or (2), then (4) follows from (1) and (3). Also, if (5) holds on the closure of $([t_0, T] \times \mathcal{R}^{2m}) \cap \mathcal{G}$, then (6) is true.

Later in this section we shall show that if F is smooth enough, then transferability is a stronger assumption than the uniform index one assumption of Subsection 3.2.2. The assumption that the nullspace depends only on t is independent of the assumption, which is made in some BDF codes [143], that the DAE is linear in the derivative, since the latter allows the nullspace to depend on x and the former allows nonlinearities in x' as long as they do not affect the nullspace of the Jacobian.

There are two important consequences of the transferability assumption. First, it makes it possible to rewrite a DAE, at least in principle, as a semi-explicit system of the same index since transferability implies that

$$F(t, x, x') = F(t, x, P(t)x'). \qquad (3.4.11)$$

Equation (3.4.11) is a purely algebraic statement concerning the variables t, x, x' in an open set and does not require $x(t)$ to be a solution of (3.4.4). Using (3.4.11), we can write (3.4.4) as

$$P(t)x' - y = 0 \qquad (3.4.12a)$$
$$F(t, x, y) = 0 \qquad (3.4.12b)$$

instead of (3.4.9). Note that (3.4.12a) implies that

$$Qy = 0 \qquad (3.4.13)$$

and that (3.4.12a) could also be written as

$$P(x' - y) - Qy = 0.$$

The key advantage of (3.4.12) is that the new larger system is still index one and the only additional solutions are the trivial algebraic solutions of $Qy = 0$. The proofs can be found in [132]. We shall give an example. Suppose that we have the index one system (3.4.4) with

$$F(t, x, x') = \begin{bmatrix} x_1' - x_1 \\ x_2 \end{bmatrix}, \quad Q = \begin{bmatrix} 0 & 0 \\ 0 & 1 \end{bmatrix}.$$

Then (3.4.9) is

$$\begin{aligned} x_1' &= y_1 \\ x_2' &= y_2 \\ 0 &= y_1 - x_1 \\ 0 &= x_2 \end{aligned}$$

which is index two. However, (3.4.12) is

$$\begin{aligned} x_1' - y_1 &= 0 \\ y_2 &= 0 \\ 0 &= y_1 - x_1 \\ 0 &= x_2 \end{aligned}$$

which is only index one.

3.4. GENERAL LINEAR MULTISTEP METHODS

The second consequence of the transferability assumption is that it enables us to obtain an explicit ODE whose solutions contain those of the DAE.

Let $C(t, u, w) = F(t, u + Q(t)w, w)$. Under the transferability assumption, $G = C_w$ is nonsingular and w is uniquely determined in terms of u, t, so that $w = w(u, t)$. Let u be the solution of the ODE

$$u' = P'u + P(I + P')w(u,t), \quad u(t_0) = P(t_0)x^0. \tag{3.4.14}$$

Then

$$x = u + Q(t)w(u, t) \tag{3.4.15}$$

is the solution of (3.4.4).

Multistep Methods

We can now give the formulation of multistep methods from [132]. Intuitively, Qx are the algebraic variables and Px the differentiated variables. An alternative formulation of (3.4.12) in terms of these new variables is

$$\begin{aligned} P[(Px)' - P'x] - Py - Qy &= 0 \\ F(t, x, y) &= 0. \end{aligned} \tag{3.4.16}$$

This formulation is more general than (3.4.4) since it only requires Px to be differentiable. Using the index one systems (3.4.12) and (3.4.16) as analogues of the semi-explicit index one system (3.4.6), we are led to two multistep formulas. The first is based on (3.4.12)

$$P_n\left\{\frac{1}{h}\sum_{j=0}^{k} a_j x_{n-j} - \sum_{j=0}^{k} b_j y_{n-j}\right\} - Q_n y_n = 0 \tag{3.4.17a}$$

$$F(t_n, x_n, y_n) = 0. \tag{3.4.17b}$$

The y in (3.4.12a) is split into Py and Qy. From (3.4.13) we have that $Qy = 0$. This is an algebraic constraint and so is merely evaluated at t_n in (3.4.17a). Notice that multiplying (3.4.17a) by Q_n gives the algebraic constraint $Q_n y_n = 0$. The term $P(t)(x'(t) - y(t))$, however is treated like (3.4.8a).

Since (3.4.17a) implies that $Q_n y_n = 0$, (3.4.17a) is the same as

$$P_n\left\{\frac{1}{h}\sum_{j=0}^{k} a_j x_{n-j} - \sum_{j=0}^{k} b_j P_{n-j} y_{n-j}\right\} - Q_n y_n = 0. \tag{3.4.18}$$

In the same way, from (3.4.16) we are led to consider the multistep methods

$$P_n\left\{\frac{1}{h}\sum_{j=0}^{k} a_j P_{n-j} x_{n-j} - \sum_{j=0}^{k} b_j [P'_{n-j} x_{n-j} + P_{n-j} y_{n-j}]\right\}$$

$$-Q_n y_n = 0 \tag{3.4.19}$$

$$F(t_n, x_n, y_n) = 0.$$

In [132] these multistep methods are considered on general nonequidistant partitions. Let π denote the partition $\{t_0 < t_1 < \cdots < t_{\tilde{n}} = T\}$ with stepsizes $h_n = t_n - t_{n-1}$. Let $h = \max_i h_i$ for a given π. Then the analysis is done over a set $\Pi = \Pi(h_{\max}, T, \nu_1, \nu_2)$ where $\pi \in \Pi$ if $h \leq h_{\max}$, $h\tilde{n} \leq (T - t_0)$ and $\nu_1 h_{i-1} \leq h_i \leq \nu_2 h_{i-1}$ for $i = 2, \ldots, \tilde{n}$. These are typical of the types of partitions generated by methods which vary the stepsize but never increase or decrease the stepsize by more than a fixed multiple (the ν_i). The variable stepsize version of the multistep method (3.4.17) is

$$P_n\left\{\frac{1}{h_n}\sum_{j=0}^{k} a_{n,j} x_{n-j} - \sum_{j=0}^{k} b_{n,j} y_{n-j}\right\} - Q_n y_n = 0 \quad (3.4.20)$$
$$F(t_n, x_n, y_n) = 0,$$

while the variable stepsize version of (3.4.19) is

$$P_n\left\{\frac{1}{h_n}\sum_{j=0}^{k} a_{n,j} P_{n-j} x_{n-j} - \sum_{j=0}^{k} b_{n,j}[P'_{n-j} x_{n-j} + P_{n-j} y_{n-j}]\right\}$$
$$-Q_n y_n = 0 \quad (3.4.21)$$
$$F(t_n, x_n, y_n) = 0.$$

This notation not only allows for variable stepsize but also for formulas of different order and type, which is the common situation in many production codes. We assume that $a_{n,0} \neq 0$.

These methods are most practical for P, Q constant, which is essentially the semi-explicit case. If P and Q are constant, then both multistep formulations are the same. For nonconstant Q, these methods can be hard to apply since they require the computation of time varying projections, or even more severely, their derivatives. However, the analysis of these methods lends insight into the behavior of numerical methods on DAE's.

To fully understand these multistep methods one needs to realize that they are, in fact, a multistep method applied to the explicit ODE (3.4.14), (3.4.15). The multistep method (3.4.19) is equivalent to an explicit multistep method applied to (3.4.14) with the added restriction necessary to keep u in the range of P. Because the multistep method (3.4.17) is actually being applied to the ODE (3.4.14), the methods acquire the correct asymptotic behavior if they have it for ODE's. This is not true for general one-leg methods.

We assume in what follows that the starting values (x_j, y_j), $j = 1, \ldots, k-1$ are known and that the DAE is transferable.

As noted earlier, BDF methods are a special type of multistep method, with $b_0 = 1$ and the other $b_i = 0$. In this case the multistep method (3.4.17) becomes

$$P_n\left\{\frac{1}{h}\sum_{j=0}^{k} a_j x_{n-j}\right\} - y_n = 0 \quad (3.4.22)$$
$$F(t_n, x_n, y_n) = 0.$$

3.4. GENERAL LINEAR MULTISTEP METHODS

However, since by (3.4.11)

$$F(t_n, x_n, y_n) = F(t_n, x_n, P_n y_n),$$

we see that (3.4.22) is algebraically equivalent to the algebraic equations (3.4.5). That is, any results on (3.4.22) will apply to (3.4.5) also. In particular, method (3.4.20) includes variable order BDF methods.

The next theorem summarizes the results on these multistep methods. A technical discussion including definitions will follow.

Theorem 3.4.1 *Suppose that (3.4.4) is transferable on $[t_0, T]$ and that $|a_{n,j}|$, $|b_{n,j}|$ are uniformly bounded for $n \geq k$ over Π. Suppose that a multistep method is stable and has order of consistency q for an ODE so that it is a qth order method. Then:*

1. *The method (3.4.21), which utilizes P', is also a qth order method for (3.4.4).*

2. *The method (3.4.20), which uses only P, has order at least one. If $q > 1$, then the order of convergence is less than q unless additional stability conditions are satisfied. However, a qth order BDF method will converge to order q.*

We now examine the statements in this theorem more carefully. Suppose that $x(t)$ is a solution of (3.4.16). For simplicity we assume that x is sufficiently smooth. Define

$$\begin{aligned}
\tau'_n &= P_n \Big\{ \frac{1}{h_n} \sum_{j=0}^{k} a_{n,j} P_{n-j} x(t_{n-j}) \\
&\quad - \sum_{j=0}^{k} b_{n,j} [P'_{n-j} x(t_{n-j}) + P_{n-j} y(t_{n-j})] \Big\} \\
\tau_n &= P_n \Big\{ \frac{1}{h_n} \sum_{j=0}^{k} a_{n,j} x(t_{n-j}) - \sum_{j=0}^{k} b_{n,j} y(t_{n-j}) \Big\}
\end{aligned}$$

so that τ'_n and τ_n are the local errors for (3.4.21) and (3.4.20) at the nth grid point of π for $n \geq k$.

The multistep method (3.4.21) is called *consistent of local order* q_n if there exists a constant c_n which is independent of stepsize such that $|\tau'_n| \leq c_n h^{q_n}$. It is *consistent of order* q if $|\tau'_n| \leq C h^q$ for all $n \geq k$. It is *consistent* if it is consistent of order at least one. A similar definition holds for τ_n.

Theorem 3.4.2 *Suppose that (3.4.4) is transferable on $[t_0, T]$. If $|a_{n,j}|$, $|b_{n,j}|$ are uniformly bounded for $n \geq k$ over Π and if*

$$\sum_{j=0}^{k} a_{n,j} = 0, \quad \sum_{j=0}^{k} a_{n,j}(t_{n-j} - t_n) = h_n \sum_{j=0}^{k} b_{n,j} \quad (3.4.23)$$

and if for $q > 1$

$$\sum_{j=0}^{k} a_{n,j}(t_{n-j} - t_n)^r = r h_n \sum_{j=0}^{k} b_{n,j}(t_{n-j} - t_n)^{r-1}, \quad 2 \leq r \leq q \quad (3.4.24)$$

then (3.4.21) has local order of consistency q.

This theorem says that the method (3.4.21) has the same order of consistency for transferable index one DAE's as for ODE's.

Theorem 3.4.3 *Under the same assumptions as Theorem 3.4.2, (3.4.20) has order of consistency q if (3.4.23) holds and if for $q > 1$, it is also true that*

$$\sum_{j=0}^{k} a_{n,j}(t_{n-j} - t_n)^r = 0, \quad \sum_{j=0}^{k} b_{n,j}(t_{n-j} - t_n)^{r-1} = 0, \quad 2 \leq r \leq q. \quad (3.4.25)$$

Observe that (3.4.25) is a stronger requirement than (3.4.24), so that in general the method (3.4.20) has a lower order of consistency for transferable DAE's than it does for ODE's. However, for BDF methods, (3.4.24) and (3.4.25) are the same so that variable stepsize variable order BDF methods have the same order of consistency for transferable DAE's as for ODE's.

Definition 3.4.2 *Let $\|\cdot\|$ denote a vector norm. Suppose that $x(t)$ is a solution of (3.4.4), and $y = (Px)' - P'x$. For each partition $\pi \in \Pi$ and arbitrary $x_n^{[i]}, y_n^{[i]}$ with $\|x_n^{[i]} - x(t_n)\| \leq \gamma$, $\|y_n^{[i]} - y(t_n)\| \leq \overline{\gamma}$ for $n = 0, \ldots, n_\pi$, $i = 1, 2$, where n_π is the number of points in π, define*

$$\alpha_n^{[i]} = P_n \left\{ \frac{1}{h_n} \sum_{j=0}^{k} a_{n,j} P_{n-j} x_{n-j}^{[i]} - \sum_{j=0}^{k} b_{n,j}[P'_{n-j} x_{n-j}^{[i]} + P_{n-j} y_{n-j}^{[i]}] \right\} - Q_n y_n^{[i]}$$

$$\beta_n^{[i]} = F(t_n, x_n^{[i]}, y_n^{[i]})$$

for $n \geq k$, $i = 1, 2$. Then the multistep method (3.4.21) is stable on Π if there are constants $\gamma, \overline{\gamma}, S$, independent of $\pi, x_n^{[i]}, y_n^{[i]}$, such that

$$\max_{n \geq k} \|x_n^{[1]} - x_n^{[2]}\| + \|y_n^{[1]} - y_n^{[2]}\|$$

$$\leq S \left\{ \max_{n \leq k-1} \{\|x_n^{[1]} - x_n^{[2]}\| + \|y_n^{[1]} - y_n^{[2]}\|\} + \max_{n \geq k} \{\|\alpha_n^{[1]} - \alpha_n^{[2]}\| + \|\beta_n^{[1]} - \beta_n^{[2]}\|\} \right\}.$$

The stability of (3.4.20) is defined analogously. Either formulation of the linear multistep methods are stable if the method is stable for explicit initial value problems. In particular, the BDF methods for $k < 7$ are stable for this class of problems for small enough h.

Previously in this chapter, we showed directly the convergence of BDF methods for uniform index one systems. Griepentrog and März [132] follow the

3.4. GENERAL LINEAR MULTISTEP METHODS

classical approach of proving stability and consistency. To see that stability and consistency imply convergence, suppose that a method is consistent of order q. Let $x_n^{[2]} = x(t_n)$, $y_n^{[2]} = y(t_n)$ be the sequence generated by the actual solution so that $\alpha_n^{[2]}, \beta_n^{[2]}$ are both $O(h^q)$. Suppose also that the initial values for the solution $x_n^{[1]}, y_n^{[1]}$ of the difference equations are $O(h^r)$ accurate. Since $\alpha_n^{[1]} = \beta_n^{[1]} = 0$ for $n \geq k$, we have that stability implies

$$\max_{n \geq k} \|x_n^{[1]} - x(t_n)\| + \|y_n^{[1]} - y(t_n)\| \leq S(O(h^r) + O(h^q)),$$

which is convergence of order $O(h^s)$ with $s = \min\{r, q\}$.

One-leg methods

The general formulation of the one-leg methods on nonequidistant partitions in [132] is

$$F\left(\bar{t}_n, \sum_{j=0}^{k} b_{n,j} x_{n-j}, \frac{1}{h_n} \sum_{j=0}^{k} a_{n,j} x_{n-j}\right) = 0 \qquad (3.4.26)$$

with $\bar{t}_n = \sum_{j=0}^{k} b_{n,j} t_{n-j}$ and $a_{n,0} \neq 0$. Again this formulation allows for a combination of formulas and type.

Assuming that $x(t)$ is a solution of (3.4.4), let

$$\tau_n = F\left(\bar{t}_n, \sum_{j=0}^{k} b_{n,j} x(t_{n-j}), \frac{1}{h_n} \sum_{j=0}^{k} a_{n,j} x(t_{n-j})\right) \qquad (3.4.27)$$

be the *local error* at the nth point of the partition π. Consistency is defined the same as for the multistep methods.

Theorem 3.4.4 *If*

$$\sum_{j=0}^{k} a_{n,j} = 0, \quad \sum_{j=0}^{k} a_{n,j}(t_{n-j} - \bar{t}_n) = h_n, \quad \sum_{j=0}^{k} b_{n,j} = 1 \qquad (3.4.28)$$

holds for (3.4.26), *then the one-leg method* (3.4.26) *is consistent for* (3.4.4). *If for $q > 1$ we also have*

$$\sum_{j=0}^{k} a_{n,j}(t_{n-j} - \bar{t}_n)^r = 0, \quad \sum_{j=0}^{k} b_{n,j}(t_{n-j} - \bar{t}_n)^{r-1} = 0 \qquad (3.4.29)$$

for $r = 2, \ldots, q$, then (3.4.26) *has order of consistency q.*

We note that (3.4.29) are the conditions such that the one-leg method has order of consistency q for an ODE. Thus the index one DAE does not add any additional order conditions. For one-leg methods, the solution of the difference equation (3.4.26) contains components in the nullspace of $F_{x'}$. Extra stability requirements are needed to insure the stability of these components. As noted earlier, the BDF methods satisfy these requirements.

The stability of one-leg methods is explicitly proven in [132] only for linear time varying systems. Suppose that we have the linear time varying problem

$$A(t)x'(t) + B(t)x(t) = g(t), \quad x(t_0) - x^0 \in \mathcal{N}(t_0). \tag{3.4.30}$$

Definition 3.4.3 *Let*

$$w_n^{[i]} = A(\bar{t}_n)\frac{1}{h_n}\sum_{j=0}^{k}a_{n,j}x_{n-j}^{[i]} + B(\bar{t}_n)\sum_{j=0}^{k}b_{n,j}x_{n-j}^{[i]} - g(\bar{t}_n).$$

Then the one-leg method (3.4.26) is stable on Π *for (3.4.30) if there is a constant* $S > 0$ *such that for arbitrary* $\pi \in \Pi$ *and arbitrary* $x_j^{[i]} \in \mathcal{R}^m$, $j = 0,\ldots,n$, $i = 1,2$, *we have*

$$\max_{n\geq k}\|x_n^{[1]} - x_n^{[2]}\| \leq S\left\{\max_{n\leq k-1}\|x_n^{[1]} - x_n^{[2]}\| + \max_{n\geq k}\|w_n^{[1]} - w_n^{[2]}\|\right\}. \tag{3.4.31}$$

A purely algebraic linear constant coefficient index one problem illustrates why extra stability conditions are required for general one-leg methods. Consider

$$x - g(t) = 0. \tag{3.4.32}$$

Applying a one-leg method to (3.4.32) gives

$$\sum_{j=0}^{k}b_j x_{n-j} - g\left(\sum_{j=0}^{k}b_j t_{n-j}\right) = 0. \tag{3.4.33}$$

For BDF methods, (3.4.33) says only that $x_n = g(t_n)$, which is the exact solution of (3.4.32). In general, however, (3.4.33) implies that if stability is to be preserved, we need at least that

$$\sigma(\lambda) = \sum_{j=0}^{k}b_j\lambda^{k-j}$$

has roots inside the closed unit disk with only simple roots of magnitude one.

However, stability rules out roots of magnitude one since they lead to a linear instability. An example is the implicit midpoint method

$$F\left(\frac{1}{2}(t_n + t_{n-1}), \frac{1}{2}(x_n + x_{n-1}), \frac{1}{h_n}(x_n - x_{n-1})\right) = 0 \tag{3.4.34}$$

which is consistent of order 2 but not stable. To see this, take the algebraic equation (3.4.32) with $g = 0$, use (3.4.34) with fixed stepsize h, and take the initial values to be $x_0^{[1]} = 0$, $x_0^{[2]} = 0$ and the perturbations to be $w_n^{[1]} =$

3.4. GENERAL LINEAR MULTISTEP METHODS

$\frac{1}{2}(-1)^n$, $w_n^{[2]} = 0$. Then $x_n^{[1]} = n(-1)^n$, $x_n^{[2]} = 0$, so that an S satisfying (3.4.31) cannot exist. Thus there is a linear growth of the errors due to the perturbations w_n. This linear instability of the implicit midpoint method, which occurs for semi-explicit index one systems, is often tolerable in practice. As long as n is not allowed to become too large, the linear growth of roundoff errors remains small compared to the truncation errors. There is a more serious form of instability for methods with roots of magnitude one applied to fully-implicit index one systems [7], which effectively rules out the practical use of these methods for this class of initial value problems (see Section 4.3 for further details).

Finally, we note that März has recently also developed characterizations of some classes of index two and index three DAE's in terms of projections [186]. BDF methods applied to semi-explicit index two systems are also considered.

Transferability and Uniform Index One

Earlier in this section, we derived results for BDF methods under the assumption of uniform index one. The results of [132] apply to a more general class of methods. However, these results apply to a somewhat different class of problems as shown by the next lemma.

Lemma 3.4.1 *Suppose that $F(t, x, x')$ is twice continuously differentiable with respect to t, x, x' and that $F(t, x, x') = 0$ is a transferable DAE. Then it is a uniform index one DAE.*

Proof. Suppose that the DAE is transferable and the partials of F have the required smoothness. We need only verify that the DAE is index one. Let $A = F_{x'}$, $B = F_x$. By assumption, $F_{x'} + F_x Q$ is nonsingular. Neither being uniform index one nor being transferable are altered by a change of variables $x = H(t)y$ with smooth nonsingular H. To show preservation of the uniform index one property, note that if $\tilde{F}(t, y, y') = F(t, Hy, H'y + Hy')$, then $\tilde{F}_{y'} = F_{x'}H$, $\tilde{F}_y = F_{x'}H' + F_y H$. It is now easy to show that the new $\tilde{F}$ system is uniform index one if and only if the F system is. Similarly transferability is not altered by such a coordinate change. Making a time varying coordinate change if necessary, we may assume that P, Q are constant and that $Px = [x_1, 0]^T$, $Qx = [0, x_2]^T$, and the DAE (3.4.4) is

$$F(t, x_1, x_2, x_1') = 0$$

where

$$F_{x'} = \begin{bmatrix} A_1 & 0 \\ A_2 & 0 \end{bmatrix}, F_x = \begin{bmatrix} B_1 & B_2 \\ B_3 & B_4 \end{bmatrix}, Q = \begin{bmatrix} 0 & 0 \\ 0 & I \end{bmatrix}.$$

Transferability then asserts that the matrix

$$\begin{bmatrix} A_1 & 0 \\ A_2 & 0 \end{bmatrix} + \begin{bmatrix} 0 & B_2 \\ 0 & B_4 \end{bmatrix}$$

is nonsingular. Since $\begin{bmatrix} A_1 \\ A_2 \end{bmatrix}$ has full column rank, there is a nonsingular matrix $\overline{P}(t, x, x')$ so that

$$\overline{P}\begin{bmatrix} A_1 \\ A_2 \end{bmatrix} = \begin{bmatrix} I \\ 0 \end{bmatrix}, \quad \overline{P}\begin{bmatrix} 0 & B_2 \\ 0 & B_4 \end{bmatrix} = \begin{bmatrix} 0 & \overline{B}_2 \\ 0 & \overline{B}_4 \end{bmatrix}$$

and $\overline{B}_4$ is nonsingular. Thus the DAE is (uniform) index one. □

We note that there are uniform index one systems which are not transferable, for example, systems which do not satisfy condition (2) of Definition 3.4.2.

Chapter 4

One-Step Methods

4.1 Introduction

In this chapter we study the order, stability and convergence properties of one-step methods when applied to the system of DAE's

$$F(t, y, y') = 0, \qquad (4.1.1)$$

where consistent initial values for $y(t_0)$ and $y'(t_0)$ are assumed to be given. One-step methods do not, in general, attain the same order of accuracy for DAE's as they do for ordinary differential equations. Of particular interest to us here are *implicit Runge-Kutta (IRK)* methods.

An M-stage IRK method applied to the DAE (4.1.1) is given by

$$F\left(t_{n-1} + c_i h, y_{n-1} + h \sum_{j=1}^{M} a_{ij} Y_j', Y_i'\right) = 0, \quad i = 1, 2, \ldots, M$$

$$y_n = y_{n-1} + h \sum_{i=1}^{M} b_i Y_i' \qquad (4.1.2)$$

where $h = t_n - t_{n-1}$. The quantities Y_i' in (4.1.2) are estimates for $y'(t_{n-1} + c_i h)$ and are called *stage derivatives*. Estimates for $y(t_{n-1} + c_i h)$ may be obtained by defining intermediate Y_i's as

$$Y_i = y_{n-1} + h \sum_{j=1}^{M} a_{ij} Y_j'. \qquad (4.1.3)$$

The method is often denoted by the shorthand notation, or *Butcher diagram*

$$\begin{array}{c|c} c & A \\ \hline & b^T \end{array} = \begin{array}{c|cccc} c_1 & a_{11} & a_{12} & \cdots & a_{1M} \\ c_2 & a_{21} & a_{22} & \cdots & a_{2M} \\ \vdots & \vdots & \vdots & \ddots & \vdots \\ c_M & a_{M1} & a_{M2} & \cdots & a_{MM} \\ \hline & b_1 & b_2 & \cdots & b_M \end{array}$$

where $\mathcal{A} = [a_{ij}]$, $b = [b_1, b_2, \ldots, b_M]^T$, and $c = [c_1, c_2, \ldots, c_M]^T$. The method (4.1.2) reduces to a standard IRK method when applied to a system of explicit ordinary differential equations (ODE's).

During the last decade, IRK methods have been the focus of increasing interest for the numerical solution of stiff ODE's. It is likely that they will play an important role in the future for the numerical solution of DAE's. One class of problems where IRK methods have an inherent potential advantage over multistep methods such as BDF is that of DAE's exhibiting frequent discontinuities. Due to their one-step nature, IRK methods are potentially more efficient for these problems than multistep methods because multistep methods must be restarted, usually at low order, after every discontinuity, whereas IRK methods can restart at a higher order. Problems with frequent discontinuities arise in several applications. For example, in the solution of partial differential equations using an adaptive mesh, each interpolation of a variable onto a new mesh generates a discontinuity of that variable in time [203]. DAE systems with frequent discontinuities also arise in the simulation of large-scale gas transmission networks [73] and in the transient analysis of electric power systems [178].

Implicit Runge-Kutta methods can also be used to generate accurate starting values for higher order BDF methods [30], thus in principle exploiting the advantages of both methods. Implicit Runge-Kutta methods, or BDF combined with IRK starters, are potentially advantageous for solving higher index nonlinear systems because they can start at a high order. Despite the convergence results for the BDF when applied to higher index systems developed in Chapter 3, existing public domain DAE software (e.g., DASSL [200] or LSODI [143]) cannot always be employed because these codes are designed to start the integration with the first order (i.e., implicit Euler) method. If the index is three or higher, this method does not converge on the first step. Even for index two systems the Newton iteration may have difficulty converging. For example, in numerical experiments described in Section 6.3 the implicit Euler method is applied to a nonlinear trajectory problem, illustrating on the first step how this method fails to converge in the algebraic variable. However, carefully chosen higher order IRK methods had no difficulty determining an accurate numerical solution in all the variables, even on the first step.

Another potential advantage of IRK methods applied to DAE's and ODE's lies in the fact that, in contrast to the case for linear multistep methods, it is possible to construct high order A-stable IRK formulas. A-stability is important when solving ODE systems having eigenvalues lying close to the imaginary axis. One application which yields this type of problem is the transient stability analysis of power systems. Thus the code of Mack [178] uses a class of A-stable one-step methods which includes higher order methods.

When solving a system with an IRK method, it is important to choose a class of methods which can be implemented efficiently. In the most general IRK method, when $\mathcal{A}$ is completely dense, there is a system of $M \cdot N$ nonlinear

4.1. INTRODUCTION

equations, where N is the size of the DAE system, which must be solved for the stage derivatives on each integration step. Compared to the expense of a multistep method, the amount of work per step must be reduced significantly before IRK methods can be competitive. Therefore, we are interested in certain classes of IRK formulas which can be implemented more efficiently than the general case. If the $\mathcal{A}$ matrix is lower triangular, the system of nonlinear equations to be solved at each step can be broken into M sets of N equations, to be solved consecutively. IRK methods of this type are said to be *semi-explicit* [194]. *Diagonally-implicit IRK* methods (*DIRK's*) are semi-explicit IRK methods with equal diagonal elements in the $\mathcal{A}$ matrix [2,67] . In this case, if one assumes the Jacobian matrix can be evaluated at one point for all the stages, then only one LU factorization is required per step since the iteration matrix is identical for each stage. Finally, if the $\mathcal{A}$ matrix has one (real) eigenvalue of multiplicity M, then Butcher [45] and Bickart [22] have shown how the IRK methods with this property can be implemented almost as efficiently as the DIRK methods. IRK methods having an $\mathcal{A}$ matrix with a single eigenvalue are referred to as *singly-implicit* IRK methods or *SIRK's* [38,194]. The term *fully-implicit* will be used to emphasize the fact that the $\mathcal{A}$ matrix does not have a special form.

There are other properties of IRK methods which will prove important during our study of their behavior on DAE's. Numerical experiments [30,34, 73,202] have led us to favor, for most classes of DAE initial value problems, methods which are *L-stable*, or even better which are *stiffly accurate*. L-stable methods perform very well when applied to index one and semi-explicit index two and index three systems, even when the initial values contain small errors. A method is *A-stable* if $\lim_{n \to \infty} y_n = 0$ for all $Re(\lambda) < 0$ and a fixed positive h when applied to the test problem $y' = \lambda y$. A method is *L-stable* if it is A-stable and, in addition, $\lim_{Re(h\lambda) \to -\infty} |y_{n+1}/y_n| = 0$. *Stiffly accurate* methods [206,236] are L-stable methods which satisfy the additional requirements that $c_M = 1$, $a_{Mj} = b_j$ for $j = 1, 2, \ldots, M$ and $\mathcal{A}$ is nonsingular.

While most of the results discussed in this chapter will pertain only to the case of a nonsingular $\mathcal{A}$ matrix, some results will be given for the special case of a singular $\mathcal{A}$ expressible in the form

$$\mathcal{A} = \begin{bmatrix} 0 & 0 \\ \underline{a} & \underline{\mathcal{A}} \end{bmatrix}, \tag{4.1.4}$$

where $\underline{\mathcal{A}}$ is nonsingular. For example, the Lobatto IIIA methods and the block implicit methods studied by Mack [178] can be expressed in this form.

The primary emphasis of this chapter will be on the theory for IRK methods. Convergence and stability theory will be developed for linear constant coefficient systems of arbitrary index in Section 4.2. The behavior of IRK methods for linear constant coefficient systems is completely understood, and illustrates many of the difficulties encountered in the solution of more complicated systems. We will also introduce some fundamental order and stability concepts in this section. In Section 4.3, we turn our attention to nonlinear index one systems, studying first the semi-explicit index one case and then the fully-implicit case. Section 4.4 gives order results for implicit Runge-Kutta methods applied to nonlinear index two systems in semi-explicit form.

There has been a significant amount of research done on IRK's applied to stiff systems which is closely related to the results given here for DAE's. In Section 4.5, we will discuss the relationship of the *stiff order* to the DAE order.

While extrapolation methods can be formulated equivalently as IRK methods, stronger results and a better intuition about these methods have so far been obtained by considering them separately and making use of their special properties. Section 4.6 will be devoted to a discussion of the recent advances in theory and practice for extrapolation methods applied to DAE's.

4.2 Linear Constant Coefficient Systems

In this section we derive necessary and sufficient conditions for the local error of an implicit Runge-Kutta method to attain a given order when applied to linear constant coefficient systems of arbitrary index ν. Then we study error propagation for constant coefficient higher index systems, and derive an expression for the global error.

Consider again the DAE system

$$F(t, y, y') = 0, \tag{4.2.1}$$

and an M-stage IRK method applied to this system,

$$F\left(t_{n-1} + c_i h, y_{n-1} + h\sum_{j=1}^{M} a_{ij} Y_j', Y_i'\right) = 0, \qquad i = 1, 2, \ldots, M,$$
$$y_n = y_{n-1} + h\sum_{i=1}^{M} b_i Y_i'. \tag{4.2.2}$$

Since y_n is determined by (4.2.2) once y_{n-1} and t_{n-1} are known, the Runge-Kutta method (4.2.2) may also be written as

$$y_n = y_{n-1} + h\psi(t_{n-1}, y_{n-1}, h). \tag{4.2.3}$$

Before proceeding, we need to define some terminology.

Definition 4.2.1 *The* local error d_n *of the method* (4.2.2) *is given by*

$$y(t_n) = y(t_{n-1}) + h\psi(t_{n-1}, y(t_{n-1}), h) - d_n. \tag{4.2.4}$$

4.2. LINEAR CONSTANT COEFFICIENT SYSTEMS

Definition 4.2.2 *The Runge-Kutta method (4.2.2) is strictly stable for the DAE (4.2.1) if the difference between a perturbed Runge-Kutta step*

$$F\left(t_{n-1} + c_i h, z_{n-1} + h\sum_{j=1}^{M} a_{ij} Z'_j + \delta_n^{(i)}, Z'_i\right) = 0, \quad 1 \le i \le M$$
$$z_n = z_{n-1} + h\sum_{i=1}^{M} b_i Z'_i + \delta_n^{(M+1)} \tag{4.2.5}$$

where $z_0 = y_0 + \delta_0^{(M+1)}$, $\|\delta_n^{(i)}\| \le \Delta$, $i = 1, 2, \ldots, M+1$, and an unperturbed Runge-Kutta step (4.2.2), satisfies $\|z_n - y_n\| \le K_0 \Delta$, where $0 < h \le h_0$, and K_0, h_0 are constants depending only on the method and the DAE.

Now consider the linear constant coefficient DAE

$$Ay' + By = f(t) \tag{4.2.6}$$

of index ν. We assume this system is solvable, so that there exist nonsingular matrices P and Q such that (see Section 2.3)

$$PAQ = \begin{bmatrix} I & 0 \\ 0 & N \end{bmatrix} \qquad PBQ = \begin{bmatrix} C & 0 \\ 0 & I \end{bmatrix}, \tag{4.2.7}$$

where I is an identity matrix and N is a nilpotent block diagonal matrix, $N = \text{diag}(N_1, N_2, \ldots, N_L)$ composed of elementary Jordan blocks of the form

$$N_i = \begin{bmatrix} 0 & & & \\ 1 & 0 & & \\ & \ddots & \ddots & \\ & & 1 & 0 \end{bmatrix}. \tag{4.2.8}$$

The matrices P and Q may be used to decouple the system (4.2.6) into an explicit ODE subsystem and a purely algebraic subsystem. A system (4.2.6) for which $PAQ = N$, $PBQ = I$, with N nilpotent, will be called a *completely singular system*. If in addition N is in the form (4.2.8) it will be called a *canonical (completely) singular system*.

Applying the IRK method to (4.2.6), we have

$$AY'_i + B\left(y_{n-1} + h\sum_{j=1}^{M} a_{ij} Y'_j\right) = f(t_{n-1} + c_i h), \quad 1 \le i \le M$$
$$y_n = y_{n-1} + h\sum_{i=1}^{M} b_i Y'_i. \tag{4.2.9}$$

Note that the coefficient matrix of the vector of stage derivatives is $A \otimes I + hB \otimes \mathcal{A}$. It can be shown that this matrix is nonsingular for small h because

$\mathcal{A}$ is nonsingular and solvability of (4.2.6) implies the pencil (A, B) is regular. Premultiplying these difference equations by P and letting $\tilde{y}_n = Q^{-1} y_n$, $\tilde{Y}'_i = Q^{-1} Y'_i$, $g(t) = Pf(t)$, we obtain

$$PAQ\tilde{Y}'_i + PBQ\left(\tilde{y}_{n-1} + h\sum_{j=1}^{M} a_{ij} \tilde{Y}'_j\right) = g(t_{n-1} + c_i h), \quad 1 \leq i \leq M$$

$$\tilde{y}_n = \tilde{y}_{n-1} + h\sum_{i=1}^{M} b_i \tilde{Y}'_i.$$

(4.2.10)

The solvability assumption implies (4.2.9) uniquely determines the Y'_i. Substituting (4.2.7) and (4.2.8) into (4.2.10) we see that the differential and singular parts of the system are decoupled. In addition, the canonical singular subsystems are decoupled from one another. Thus it is sufficient to study the behavior of the IRK method on a *canonical singular subsystem* and an explicit ODE in order to understand its behavior on general linear constant coefficient systems.

Consider a canonical singular subsystem of index ν

$$Ny' + y = g(t) \tag{4.2.11}$$

where N is a $\nu \times \nu$ matrix of the form (4.2.8), $g(t) = [g_1(t), g_2(t), \ldots, g_\nu(t)]^T$, and $y(t) = [y_1(t), y_2(t), \ldots, y_\nu(t)]^T$. The solution to (4.2.11) is given by

$$\begin{aligned} y_1(t) &= g_1(t) \\ y_2(t) &= g_2(t) - g'_1(t) \\ &\vdots \\ y_\nu(t) &= g_\nu(t) + \sum_{i=1}^{\nu-1} (-1)^{\nu-i} g_i^{(\nu-i)}(t). \end{aligned}$$

Applying the IRK method to (4.2.11), we obtain

$$NY'_i + y_{n-1} + h\sum_{j=1}^{M} a_{ij} Y'_j = g(t_{n-1} + c_i h), \quad 1 \leq i \leq M$$

$$y_n = y_{n-1} + h\sum_{i=1}^{M} b_i Y'_i.$$

(4.2.12)

For the important case of a nonsingular $\mathcal{A}$ matrix let $Y'_i = [Y'_{i,1}, Y'_{i,2}, \ldots, Y'_{i,\nu}]^T$, where $Y'_{i,j}$ denotes the ith stage derivative corresponding to the jth component of the solution vector. As noted, the difference equations (4.2.12) can be solved uniquely for the stage derivatives $Y' = [Y'_1, Y'_2, \ldots, Y'_M]^T$. The structure of N causes the solution to the ith equation in the original system (4.2.11) to depend only on the solutions to the first $(i-1)$ equations. A similar dependency is present in the difference equations (4.2.12), allowing us to solve first for

4.2. LINEAR CONSTANT COEFFICIENT SYSTEMS

the stage derivatives $Y'_{1,1}, Y'_{2,1}, \ldots, Y'_{M,1}$ corresponding to $y_1(t)$, second for the stage derivatives $Y'_{1,2}, Y'_{2,2}, \ldots, Y'_{M,2}$ corresponding $y_2(t)$, etc.

In general, for each component of the solution $y_j(t)$, $j = 1, 2, \ldots, \nu$, we solve a subset of M equations from system (4.2.12) for the corresponding stage derivatives

$$[Y'_{1,j}, Y'_{2,j}, \ldots, Y'_{M,j}]^T =$$
$$\frac{1}{h} A^{-1} \left([g_j(t_{n-1} + c_1 h), g_j(t_{n-1} + c_2 h), \ldots, g_j(t_{n-1} + c_M h)]^T \right.$$
$$\left. - y_{n-1,j} \epsilon_M - [Y'_{1,j-1}, Y'_{2,j-1}, \ldots, Y'_{M,j-1}]^T \right) \quad (4.2.13)$$

where $\epsilon_M = [1, 1, \ldots, 1]^T$. Note that the stage derivatives depend only on the numerical solution $y_{n-1,j}$ and on the stage derivatives of the $(j-1)$st variable. Define

$$G_i = \frac{1}{h} \begin{bmatrix} g_i(t_{n-1} + c_1 h) - g_i(t_{n-1}) \\ g_i(t_{n-1} + c_2 h) - g_i(t_{n-1}) \\ \vdots \\ g_i(t_{n-1} + c_M h) - g_i(t_{n-1}) \end{bmatrix}$$

$$c^i = [c^i_1, c^i_2, \ldots, c^i_M]^T, \quad \text{where } c^i_j = (c_j)^i.$$

The local stage derivatives are the stage derivatives computed under the assumption that the value of y_{n-1} is exact. Making the local error assumption,

$$y_{n-1} = y(t_{n-1}) \quad (4.2.14)$$

and substituting for the $(j-1)$st stage derivatives in (4.2.13), we find the following expressions for the *local* stage derivatives:

$$\left[Y'_{1,1}, Y'_{2,1}, \ldots, Y'_{M,1}\right]^T = A^{-1} G_1$$

$$\left[Y'_{1,2}, Y'_{2,2}, \ldots, Y'_{M,2}\right]^T = A^{-1} G_2 - \frac{1}{h} A^{-2} G_1 + \frac{1}{h} A^{-1} \epsilon_M g'_1(t_{n-1})$$

$$\left[Y'_{1,3}, Y'_{2,3}, \ldots, Y'_{M,3}\right]^T = A^{-1} G_3 + \frac{1}{h} A^{-1} \epsilon_M [g'_2(t_{n-1}) - g''_1(t_{n-1})]$$
$$- \frac{1}{h} A^{-2} G_2 + \frac{1}{h^2} A^{-3} G_1$$
$$- \frac{1}{h^2} A^{-2} \epsilon_M g'_1(t_{n-1}) \quad (4.2.15)$$

and similar expressions for the remaining components. The *local error* d_n is given by

$$d_n = y(t_{n-1}) + h \sum_{i=1}^{M} b_i Y'_i - y(t_n) \quad (4.2.16)$$

where $d_n = [d_{n,1}, d_{n,2}, \ldots, d_{n,\nu}]^T$ and Y'_i represent the local stage derivatives given by (4.2.15). Expanding (4.2.16) in a Taylor series about t_{n-1}, and equating like powers of h, it is easy to see that the local error $d_{n,1}$ in the first component satisfies

$$d_{n,1} = O(h^{k_{a,1}+1}), \qquad (4.2.17)$$

when

$$b^T \mathcal{A}^{-1} c^i = 1, \qquad i = 1, 2, \ldots, k_{a,1}. \qquad (4.2.18)$$

The integer $k_{a,1}$ in (4.2.18) will be called the *algebraic order* of the IRK method applied to index one constant coefficient canonical algebraic systems. Note that for stiffly accurate methods, $b_i = a_{Mi}$, $i = 1, 2, \ldots, M$ and $c_M = 1$, so that $k_{a,1} = \infty$. In other words, stiffly accurate methods are exact on this class of problems. For the second component, the local error $d_{n,2}$ is given by

$$\begin{aligned} d_{n,2} &= y_2(t_{n-1}) - y_2(t_n) + h b^T \mathcal{A}^{-1} G_2 \\ &\quad + b^T \mathcal{A}^{-1} \epsilon_M g_1'(t_{n-1}) - b^T \mathcal{A}^{-2} G_1. \end{aligned} \qquad (4.2.19)$$

Note that $g_2(t_{n-1}) - g_2(t_n) + h b^T \mathcal{A}^{-1} G_2 = O(h^{k_{a,1}+1})$ if we assume the IRK method has algebraic order $k_{a,1}$ on index one problems. Then expand the remaining terms in (4.2.19) in a Taylor series about t_{n-1} and equate like powers of h to obtain the following set of order conditions for index two constant coefficient systems,

$$\begin{aligned} b^T \mathcal{A}^{-1} \epsilon_M &= b^T \mathcal{A}^{-2} c \\ b^T \mathcal{A}^{-2} c^i &= i, \qquad i = 2, 3, \ldots, k_{a,2}. \end{aligned}$$

We define the algebraic order of the IRK method applied to index two constant coefficient canonical algebraic systems to be $k_{a,2}$ if these conditions are satisfied. The local error for a general index two constant coefficient system thus satisfies $d_{n,2} = O(h^{k_{a,2}}) + O(h^{k_{a,1}+1})$.

The analysis can be extended in a straightforward manner to the general index ν case. Thus we have shown,

Proposition 4.2.1 *For a completely singular linear constant coefficient system of index ν, the local error satisfies*

$$d_{n,\nu} = O(h^{k_{a,\nu}-\nu+2}) + O(h^{k_{a,\nu-1}-\nu+3}) + \cdots + O(h^{k_{a,1}+1}) \qquad (4.2.20)$$

where $k_{a,\nu}$ is the largest integer such that

$$\begin{aligned} b^T \mathcal{A}^{-i} \epsilon_M &= \frac{b^T \mathcal{A}^{-\nu} c^{\nu-i}}{(\nu - i)!}, \qquad i = 1, 2, \ldots, \nu - 1 \\ b^T \mathcal{A}^{-\nu} c^i &= i(i-1) \cdots (i - \nu + 1), \qquad i = \nu, \nu+1, \ldots, k_{a,\nu}. \end{aligned}$$

Clearly, the higher the index the more difficult it is to find IRK methods which are convergent in all the variables, even for linear constant coefficient systems. Finding an IRK method having the same rate of convergence in all of the variables poses severe restrictions on the coefficients.

4.2. LINEAR CONSTANT COEFFICIENT SYSTEMS

The preceding analysis can also be carried through when the IRK method has a singular $\mathcal{A}$ matrix in the form (4.1.4). The best known of such methods is the implicit trapezoid method. Other examples include the family of Lobatto IIIA methods and the block implicit one-step methods analyzed in [240]. To implement these methods for the fully-implicit DAE (4.1.1), it is necessary to assume that initial values for the derivatives of all the variables are given. Hence $Y_1' = y'$ initially. At the end of each step, we set $Y_{1,n+1}' = Y_{M,n}'$. Thus, rather than solving for all the stage derivatives Y' as we did before in (4.2.15), we now solve (4.2.12) for $\underline{Y}' = [Y_2', Y_3', \ldots, Y_M']$ using the fact that $\underline{\mathcal{A}}$ is nonsingular. It is then straightforward to derive conditions which ensure the local error in the index one variable satisfies (4.2.17) if and only if

$$\begin{aligned} \underline{b}^T \underline{\mathcal{A}}^{-1} \underline{a} &= b_1 \\ \underline{b}^T \underline{\mathcal{A}}^{-1} \underline{c}^j &= 1, \qquad j = 1, 2, \ldots, k_{a,1} \end{aligned} \qquad (4.2.21)$$

where $\underline{b} = [b_2, b_3, \ldots, b_M]^T$ and $\underline{c} = [c_2, c_3, \ldots, c_M]^T$. The Lobatto IIIA and the block implicit one-step methods have the additional property that $b_i = a_{Mi}, i = 1, 2, \ldots, M$ so that $\underline{b}^T \underline{\mathcal{A}}^{-1} = [0, \ldots, 0, 1]$. Thus for these particular methods, $k_{a,1} = \infty$. That is, if the initial values for both y and y' contain no inconsistencies, then these methods will determine the numerical solution exactly when applied to the simple algebraic equation $y(t) = g(t)$. However, note that $\lim_{Re(z) \to -\infty} |y_{n+1}/y_n| = 1$ so there is no dampening of any perturbations due to initial or roundoff errors.

Next we examine the propagation of errors for IRK methods applied to linear constant coefficient systems. Solving (4.2.11) by the perturbed Runge-Kutta method (4.2.5), we have

$$NZ_i' + z_{n-1} + h\sum_{j=1}^{M} a_{ij} Z_j' - \delta_n^{(i)} = g(t_{n-1} + c_i h), \quad 1 \leq i \leq M$$

$$z_n = z_{n-1} + h\sum_{i=1}^{M} b_i Z_i' - \delta_n^{(M+1)},$$
(4.2.22)

where the perturbations $\delta_n^{(i)} = [\delta_{n,1}^{(i)}, \delta_{n,2}^{(i)}, \ldots, \delta_{n,\nu}^{(i)}]^T$ could be due to roundoff error, errors in solving the linear systems at each stage, or could be interpreted as truncation errors at each stage. Subtracting (4.2.22) from (4.2.12) and defining $e_n = y_n - z_n$, $E_i' = Y_i' - Z_i'$, we obtain an expression for the difference between these two solutions

$$NE_i' + e_{n-1} + h\sum_{j=1}^{M} a_{ij} E_j' + \delta_n^{(i)} = 0, \quad 1 \leq i \leq M$$

$$e_n = e_{n-1} + h\sum_{i=1}^{M} b_i E_i' + \delta_n^{(M+1)}.$$
(4.2.23)

By solving the first equation in (4.2.23) for E_i' and substituting into the second equation, we can obtain a relation describing the error propagation of the method. For linear constant coefficient index one systems we then have

$$e_{n,1} = (1 - b^T A^{-1} \epsilon_M) e_{n-1,1} - (b^T A^{-1} \delta_{n,1} - \delta_{n,1}^{(M+1)}), \qquad (4.2.24)$$

where $\delta_{n,j} = [\delta_{n,j}^{(1)}, \delta_{n,j}^{(2)}, \ldots, \delta_{n,j}^{(M)}]^T$. Thus we have shown,

Theorem 4.2.1 *An implicit Runge-Kutta method (4.2.2) is strictly stable for linear constant coefficient index one DAE's iff the method coefficients satisfy*

$$|1 - b^T \mathcal{A}^{-1} \epsilon_M| < 1. \tag{4.2.25}$$

Defining the *stability constant* r by $r = 1 - b^T \mathcal{A}^{-1} \epsilon_M$, we will say that an IRK method satisfies the *strict stability condition* if $|r| < 1$, and that it satisfies the *stability condition* if $|r| \leq 1$.

For higher index systems, the error propagation relation for the index two variable is given by

$$\begin{aligned} e_{n,2} &= (1 - b^T \mathcal{A}^{-1} \epsilon_M) e_{n-1,2} - (b^T \mathcal{A}^{-1} \delta_{n,2} - \delta_{n,2}^{(M+1)}) \\ &\quad + \frac{1}{h} b^T \mathcal{A}^{-2} (\delta_{n,1} + \epsilon_M e_{n-1,1}). \end{aligned} \tag{4.2.26}$$

For the general index ν case, the error propagation relation is

$$\begin{aligned} e_{n,\nu} &= (1 - b^T \mathcal{A}^{-1} \epsilon_M) e_{n-1,\nu} - (b^T \mathcal{A}^{-1} \delta_{n,\nu} - \delta_{n,\nu}^{(M+1)}) \\ &\quad - \sum_{i=1}^{\nu-1} \frac{(-1)^i}{h^i} b^T \mathcal{A}^{-i-1} (\delta_{n,\nu-i} + \epsilon_M e_{n-1,\nu-i}). \end{aligned} \tag{4.2.27}$$

Note that the strict stability condition is no longer sufficient to ensure stability, in a strict mathematical sense, of the IRK method when applied to linear constant coefficient systems of index greater than one. For small stepsizes, roundoff errors can be significant in the solution components which occur in higher index systems because of the negative powers of h in (4.2.27). On the other hand, the strict stability condition does ensure that errors do not *accumulate* in a disastrous way.

For the index one case, it is simple to estimate the size of the global error now that we have an understanding of the local truncation error and the stability properties of the method. Let $\delta_{n,1}$ and $\delta_{n,1}^{(M+1)}$ represent the internal (i.e, of each stage) local truncation errors. Then in the absence of roundoff errors, and under the local error assumption (4.2.14), it follows from (4.2.24) that

$$e_{n,1} = -(b^T \mathcal{A}^{-1} \delta_{n,1} - \delta_{n,1}^{(M+1)}).$$

Since the local error $d_{n,1}$ has been shown to be of order $O(h^{k_{a,1}+1})$, and since under the local error assumption $e_{n,1} = d_{n,1}$, we have

$$-(b^T \mathcal{A}^{-1} \delta_{n,1} - \delta_{n,1}^{(M+1)}) = O(h^{k_{a,1}+1}).$$

Next we rewrite the expression for the global error given in equation (4.2.24) as

$$e_{n,1} = r^n e_{0,1} - \sum_{i=0}^{n-1} r^i \left(b^T \mathcal{A}^{-1} \delta_{n-i,1} - \delta_{n-i,1}^{(M+1)} \right). \tag{4.2.28}$$

4.2. LINEAR CONSTANT COEFFICIENT SYSTEMS

Note that if $r = 0$ in (4.2.28), then the global error is equal to the local error. For $|r| < 1$, it follows from (4.2.28) that

$$\|e_{n,1}\| \leq r^n \|e_0\| + \left(\frac{1 - r^n}{1 - r}\right) M h^{k_{a,1}+1}$$

for some positive constant M. Since r is independent of h and $\lim_{n\to\infty}(1 - r^n)/(1 - r) = 1/(1 - r)$, we have in this case that the order of the global error is the same as the order of the local error.

IRK methods can be useful for the solution of higher index systems, provided that we understand the implications of the error propagation relations (4.2.26) and (4.2.27) given above. We can see that the sensitivity to roundoff errors is confined to the later components of the system, which exhibit the higher index behavior, and does not propagate back into the earlier components. Finally, using the error propagation relations (4.2.27), we can extend the conclusions for global error in solving linear constant coefficient index one systems to higher index systems as follows.

Definition 4.2.3 *The* constant coefficient order *of an IRK method* (4.2.2) *is equal to* $k_{c,\nu}$ *if the method converges with global error* $O(h^{k_{c,\nu}})$ *for all solvable linear constant coefficient systems* (4.2.6) *of index* $\leq \nu$.

Theorem 4.2.2 *Suppose the IRK method* (4.2.2) *satisfies the strict stability condition and has a nonsingular A matrix. Then the constant coefficient order $k_{c,\nu}$ of the global error of this method is given by*

$$k_{c,\nu} = \min_{1 \leq i \leq \nu} (k_d, k_{a,i} - \nu + 2) \qquad (4.2.29)$$

where k_d is the order of the method for purely differential (nonstiff) systems.

It has been noted by März [181] that the application of the implicit midpoint rule to the simple algebraic equation $y(t) = 0$ can be unstable if small perturbations are introduced into the difference equations. Let us consider for the moment relaxing the strict stability condition so that $r = \pm 1$. Note that $r = -1$ for the implicit midpoint rule. Then for linear constant coefficient index one systems, we see from equation (4.2.28) that if $|r| = 1$, then in the worst case there is a linear growth of the perturbations (i.e., $O(n\delta)$). If the perturbations represent the internal local truncation errors, then we would expect that the constant coefficient order will satisfy $k_{c,1} = \min(k_d, k_{a,1})$ if $r = 1$ and $k_{c,1} = \min(k_d, k_{a,1} + 1)$ if $r = -1$, where in the latter case a cancellation of errors between steps allows a higher order of accuracy than first expected. In fact, we see that as long as n is not allowed to become too large, then the linear growth of roundoff errors will remain small with respect to the accumulation of local truncation errors. Thus, we would expect the implicit midpoint rule to remain second order accurate when the stepsize is bounded away from zero.

Finally, we present some results on the order of accuracy of IRK methods from the stiff ODE literature applied to index one and index two linear constant coefficient systems. The L-stable methods we have chosen to investigate here are:

1. 2-stage, '2nd order' singly-implicit method (SIRK) [39], with $\lambda = 1 - \sqrt{2}/2$

2. 5-stage, '4th order' diagonally-implicit method (DIRK) [2,67]

3. 3-stage, '3rd order' singly-implicit method (SIRK) [39], with $1/\lambda$ the root of the Laguerre polynomial of degree three

4. 7-stage, '3rd order' extrapolation method based on the implicit Euler method and polynomial extrapolation.

Methods (1) and (2) are stiffly accurate. The results are given in Table 4.2.1, where it can be seen that, as observed above, it is difficult to maintain the same rate of convergence in all of the variables for linear constant coefficient index two systems.

L-Stable Methods	ODE Order k_d	Index 1 Order $k_{a,1}$	Index 2 Order $k_{a,2}$
1. Two-stage SIRK	2	∞	2
2. Five-stage DIRK	4	∞	1
3. Three-stage SIRK	3	3	2
4. Seven-stage Extrp.	3	∞	3

Table 4.2.1: Order of Consistency for Linear Constant Coefficient Systems

4.3 Nonlinear Index One Systems

4.3.1 Semi-Explicit Index One Systems

An important class of DAE's which arise frequently in practice are semi-explicit index one systems

$$\begin{aligned} x' &= f(x, y, t) \\ 0 &= g(x, y, t) \end{aligned} \quad (4.3.1)$$

where $(\partial g/\partial y)^{-1}$ exists and is bounded in a neighborhood of the exact solution. These systems have been the most frequently studied class of nonlinear DAE's, and consequently are the best understood. In this form, the variables are separated into two types: y variables which appear in the system in only an algebraic way and x variables for which explicit differential equations are given. Consequently, there is more than one way to implement an IRK method

4.3. NONLINEAR INDEX ONE SYSTEMS

in order to take advantage of this system structure. The most natural implementation follows from the traditional ODE application of these methods to the singular perturbation problem,

$$\begin{aligned} x' &= f(x,y,t) \\ \epsilon y' &= g(x,y,t) \end{aligned} \qquad (4.3.2)$$

in the limiting case that ϵ is set to zero. The set of difference equations which result are

$$X'_i = f\left(x_{n-1} + h\sum_{j=1}^{M} a_{ij}X'_j, y_{n-1} + h\sum_{j=1}^{M} a_{ij}Y'_j, t_{n-1} + c_i h\right) \quad (4.3.3a)$$

$$0 = g\left(x_{n-1} + h\sum_{j=1}^{M} a_{ij}X'_j, y_{n-1} + h\sum_{j=1}^{M} a_{ij}Y'_j, t_{n-1} + c_i h\right) \quad (4.3.3b)$$

for $i = 1, 2, \ldots, M$. Solving for the stage derivatives X'_i, Y'_i for $i = 1, 2, \ldots, M$, the numerical solution is advanced as usual

$$\begin{aligned} x_n &= x_{n-1} + h\sum_{i=1}^{M} b_i X'_i \\ y_n &= y_{n-1} + h\sum_{i=1}^{M} b_i Y'_i. \end{aligned} \qquad (4.3.4)$$

The definition of an IRK method for a fully-implicit DAE given in (4.1.2) leads to the same system of difference equations.

It is easy to see that IRK methods applied in this way to semi-explicit index one DAEs achieve the same order of accuracy in the x variable for the DAE as for standard nonstiff ODEs. This follows simply because by the implicit function theorem, g can be solved for the stage approximations Y_i in terms of X_i and t_i, where

$$\begin{aligned} X_i &= x_{n-1} + h\sum_{j=1}^{M} a_{ij}X'_j \\ Y_i &= y_{n-1} + h\sum_{j=1}^{M} a_{ij}Y'_j \end{aligned}$$

and $t_i = t_{n-1} + c_i h$, and inserted into (4.3.3a) to yield an equation which is the IRK method applied to the underlying ODE.

In general, the numerical solution does not automatically satisfy the constraint

$$g(x_n, y_n, t_n) = 0. \qquad (4.3.5)$$

However, the solutions to stiffly accurate methods always satisfy the constraint. These methods have the property that $x_n = X_M$ and $y_n = Y_M$.

By the implicit function theorem, g can be solved for y_n in terms of x_n and t_n. Thus for stiffly accurate methods, y_n also attains the ODE order of accuracy. This result was originally noted by Deuflhard et al.[96] and by Griepentrog and März [132]. We also note that if the IRK method is stiffly accurate, (4.3.3) can be solved directly for Y_i, rather than for Y_i'. This formulation is also applicable to the Lobatto IIIA methods and the block implicit one-step methods of [240].

For IRK methods whose solutions do not automatically satisfy the constraint (4.3.5), it is possible to force the constraint to be satisfied by computing y_n as the solution to (4.3.5), given x_n computed by the IRK method. This technique has been utilized frequently in practice (see for example Cameron [48]), and ensures that the numerical solution will be computed to the expected ODE order of accuracy. A disadvantage is that the extra nonlinear system increases the expense of the method.

For IRK methods of the form (4.3.3) which are not stiffly accurate and where the constraint (4.3.5) has not been enforced at t_n, the order results given later in this section for fully-implicit index one systems apply, but they are not sharp for some methods. For example, extrapolation methods based on the implicit Euler method or on the implicit midpoint method achieve an order greater than these bounds (see Section 4.6). Generalizing the Butcher tree theory for ODE's, Roche [214] has derived a set of necessary and sufficient order conditions for IRK methods satisfying the strict stability condition applied to semi-explicit index one DAE's. These conditions include the order conditions derived earlier for index one linear constant coefficient systems, as well as additional conditions which must be satisfied to achieve an order greater than two.

Convergence results for methods which satisfy the stability condition, but not the strict stability condition, are given for semi-explicit index one systems by Ascher [6] and Ascher and Bader [8]. In Ascher [6], a convergence result is presented for collocation schemes which are applied in such a way that the algebraic components of the system are approximated in a piecewise discontinuous space. Since the differential components are generally one derivative smoother than the algebraic components in a semi-explicit index one DAE, it is natural that the numerical approximations should replicate this property. This separation in the treatment of the differential and algebraic parts of the system is recommended when, as in semi-explicit index one systems, it can be conveniently achieved.

4.3.2 Fully-Implicit Index One Systems

This subsection considers the order, stability and convergence properties of IRK methods applied to fully-implicit index one systems of DAE's

$$F(t, y, y') = 0, \qquad (4.3.6)$$

4.3. NONLINEAR INDEX ONE SYSTEMS

where consistent initial values for $y(t_0)$ and $y'(t_0)$ are assumed to be given. In general, IRK methods do not attain the same order of accuracy for fully-implicit index one systems as they do for systems with a simpler mathematical structure (e.g., such as linear constant coefficient or semi-explicit systems). This additional loss of accuracy is due to time dependent mixing which can occur between the errors in the differential and algebraic parts of the system.

Before we can state the main result of this subsection, which is the order conditions for IRK methods applied to index one systems, we need to make a few definitions.

Definition 4.3.1 *The ith internal local truncation error $\delta_i^{(n)}$ at t_n of an M-stage IRK method (4.1.2) is given by*

$$\delta_i^{(n)} = y(t_{n-1}) + h\sum_{j=1}^{M} a_{ij} y'(t_{n-1} + c_j h) - y(t_{n-1} + c_i h) \qquad i = 1, 2, \ldots, M$$

$$\delta_{M+1}^{(n)} = y(t_{n-1}) + h\sum_{i=1}^{M} b_i y'(t_{n-1} + c_i h) - y(t_n). \qquad (4.3.7)$$

Definition 4.3.2 *The internal stage order k_I of an M-stage IRK method (4.1.2) is given by*

$$k_I = \min(k_1, \ldots, k_M, k_{M+1})$$

where

$$\delta_i = O(h^{k_i+1}), \qquad i = 1, \ldots, (M+1).$$

Defining the algebraic conditions $C(q)$, $B(q)$, $A_1(q)$ by

$$\begin{aligned}
C(q) &: \sum_{j=1}^{M} a_{ij}\, c_j^{k-1} = c_i^k/k, \qquad i = 1, \ldots, M, \quad k = 1, \ldots, q \\
B(q) &: \sum_{j=1}^{M} b_j\, c_j^{k-1} = 1/k, \qquad k = 1, \ldots, q \\
A_1(q) &: b^T \mathcal{A}^{-1} c^k = 1, \qquad k = 1, \ldots, q
\end{aligned} \qquad (4.3.8)$$

it is easy to see, by expanding (4.3.7) in Taylor series around t_{n-1}, as in [105,106], that an IRK method has internal stage order k_I iff $C(k_I)$ and $B(k_I)$ hold. Furthermore, if $C(q)$ and $B(q+1)$ hold, then the classical (nonstiff) ODE order k_d is at least $q+1$. The definition of $A_1(q)$ corresponds to the order conditions for index one constant coefficient systems (i.e., $k_{a,1} = q$ iff $A_1(q)$). In general, for IRK methods with nonsingular $\mathcal{A}$ matrices, $B(q)$ together with $C(q)$ implies $A_1(q)$.

Finally, to prove the main result we will need the following lemma, which is due to A. C. Hindmarsh. The proof in [42] requires modification if $r = -1$.

Lemma 4.3.1 *Let*

$$K = \begin{bmatrix} I_{m_1} & 0 \\ 0 & rI_{m_2} \end{bmatrix}, \quad -1 \leq r < 1. \tag{4.3.9}$$

Then

$$(K + O(h))^n = \begin{bmatrix} O(1) & O(h) \\ O(h) & |r|^n O(1) + O(h) \end{bmatrix},$$

where $O(h^\delta)$, $\delta = 0, 1$, denotes a matrix whose elements are all bounded by a constant (which is independent of n) times h^δ, and $n \leq 1/h$.

Now we can state the main result of this subsection, which gives conditions on the coefficients of a IRK method so that it attains a given order of accuracy when applied to fully-implicit uniform index one (see Definition 3.2.1) DAE systems. An outline of the proof is as follows. First a recurrence relation which describes the propagation of errors of the system is derived. Then a time-dependent transformation into Kronecker canonical form, which is possible by Lemma 3.2.1, is made. With this change of coordinates, the stability matrix which propagates the errors is K. Then Lemma 4.3.1 is used to bound the powers of K, and the recurrence relations are solved for the errors. Finally, the higher order nonlinear terms which have been neglected are shown to be small.

Theorem 4.3.1 *Suppose (4.3.6) is uniform index one, the IRK method satisfies the stability condition $|r| \leq 1$, the errors in the initial conditions are $O(h^G)$, and the errors in terminating the Newton iterations are $O(h^{G+\delta})$ where $\delta = 1$ if $|r| = 1$ and $\delta = 0$ otherwise, and $G > 1$. Then the global errors satisfy $||e_n|| = O(h^G)$ where*

$$G = \begin{cases} q, & \text{if } C(q) \text{ and } B(q) \\ q+1, & \text{if } C(q),\ B(q+1) \text{ and } -1 \leq r < 1 \\ q+1, & \text{if } C(q),\ B(q+1),\ A_1(q+1) \text{ and } r = 1 \end{cases}$$

Proof. Consider the IRK method (4.1.2). The numerical solution satisfies

$$F\left(t_{n-1} + c_i h, y_{n-1} + h \sum_{j=1}^M a_{ij} Y_j', Y_i'\right) = \bar{\eta}^{(i)}, \quad 1 \leq i \leq M$$

$$y_n = y_{n-1} + h \sum_{i=1}^M b_i Y_i' \tag{4.3.10}$$

where $\bar{\eta}^{(i)}$, $i = 1, 2, \ldots, M$ represent errors from terminating the Newton iteration. The true solution satisfies

$$F\left(t_{n-1} + c_i h,\ y(t_{n-1}) + h \sum_{j=1}^M a_{ij} y'(t_{n-1} + c_j h) - \delta_i,\ y'(t_{n-1} + c_i h)\right) = 0$$

$$y(t_n) = y(t_{n-1}) + h \sum_{i=1}^M b_i y'(t_{n-1} + c_i h) - \delta_{M+1},$$

$$\tag{4.3.11}$$

4.3. NONLINEAR INDEX ONE SYSTEMS

where δ_i, $i = 1, 2, \ldots, M + 1$ are the internal local truncation errors at t_n. Let $E'_i = Y'_i - y'(t_{n-1} + c_i h)$, $E_i = Y_i - y(t_{n-1} + c_i h)$, and $e_n = y_n - y(t_n)$. Subtracting (4.3.11) from (4.3.10), we obtain

$$A_i E'_i + B_i \left(e_{n-1} + h \sum_{j=1}^{M} a_{ij} E'_j + \delta_i \right) = \eta_i, \quad i = 1, 2, \cdots, M, \quad (4.3.12a)$$

$$e_n = e_{n-1} + h \sum_{i=1}^{M} b_i E'_i + \delta_{M+1}, \quad (4.3.12b)$$

where $A_i = \partial F / \partial y'$ and $B_i = \partial F / \partial y$ are evaluated at $(t_{n-1} + c_i h, y(t_{n-1} + c_i h), y'(t_{n-1} + c_i h))$ and η_i is the sum of residuals from the Newton iteration and higher order terms in e_{n-1} and E'_i.

We can rewrite (4.3.12a) in the form

$$\begin{bmatrix} A_1 + a_{11} h B_1 & a_{12} h B_1 & \cdots & a_{1M} h B_1 \\ a_{21} h B_2 & A_2 + a_{22} h B_2 & \cdots & a_{2M} h B_2 \\ \vdots & \vdots & \ddots & \vdots \\ a_{M1} h B_M & a_{M2} h B_M & \cdots & A_M + a_{MM} h B_M \end{bmatrix} \begin{bmatrix} E'_1 \\ E'_2 \\ \vdots \\ E'_M \end{bmatrix}$$

$$= - \begin{bmatrix} B_1(e_{n-1} + \delta_1) \\ B_2(e_{n-1} + \delta_2) \\ \vdots \\ B_M(e_{n-1} + \delta_M) \end{bmatrix} + \begin{bmatrix} \eta_1 \\ \eta_2 \\ \vdots \\ \eta_M \end{bmatrix}. \quad (4.3.13)$$

For notational convenience, we will henceforth assume that all matrices without subscripts or superscripts are evaluated at $(t_n, y(t_n), y'(t_n))$. Let $A = \partial F / \partial y'$ and $B = \partial F / \partial y$ be evaluated at $(t_n, y(t_n), y'(t_n))$, and let P and Q be transformation matrices which bring A and B to Kronecker canonical form, as in Lemma 3.2.1. Let $\tilde{e}_{n-1} = Q^{-1} e_{n-1}$, $\tilde{E}'_i = Q^{-1} E'_i$, $\tilde{\delta}_i = Q^{-1} \delta_i$, and $\tilde{\eta}_i = P_i \eta_i$. Then we can rewrite (4.3.13),

$$\begin{bmatrix} X_1 + a_{11} h W_1 & a_{12} h W_1 & \cdots & a_{1M} h W_1 \\ a_{21} h W_2 & X_2 + a_{22} h W_2 & \cdots & a_{2M} h W_2 \\ \vdots & \vdots & \ddots & \vdots \\ a_{M1} h W_M & a_{M2} h W_M & \cdots & X_M + a_{MM} h W_M \end{bmatrix} \begin{bmatrix} \tilde{E}'_1 \\ \tilde{E}'_2 \\ \vdots \\ \tilde{E}'_M \end{bmatrix}$$

$$= - \begin{bmatrix} W_1(\tilde{e}_{n-1} + \tilde{\delta}_1) \\ W_2(\tilde{e}_{n-1} + \tilde{\delta}_2) \\ \vdots \\ W_M(\tilde{e}_{n-1} + \tilde{\delta}_M) \end{bmatrix} + \begin{bmatrix} \tilde{\eta}_1 \\ \tilde{\eta}_2 \\ \vdots \\ \tilde{\eta}_M \end{bmatrix}, \quad (4.3.14)$$

where $W_i = P_i B_i Q_i (Q_i^{-1} Q)$ and $X_i = P_i A_i Q_i (Q_i^{-1} Q)$.

92   CHAPTER 4. ONE-STEP METHODS

By the definition of P, Q and the smoothness of Q^{-1} and Q (Lemma 3.2.1), we have

$$W_i = \begin{bmatrix} C_i & 0 \\ 0 & I_{m_2} \end{bmatrix}(I+O(h)),$$

$$X_i = \begin{bmatrix} I_{m_1} & 0 \\ 0 & 0 \end{bmatrix}(I+O(h)). \quad (4.3.15)$$

Partition $\tilde{E}'_i$ into $\tilde{E}'^{(1)}_i$ and $\tilde{E}'^{(2)}_i$, where $\tilde{E}'^{(1)}_i$ has dimension m_1 and $\tilde{E}'^{(2)}_i$ has dimension m_2. By partitioning $\tilde{\delta}_i$, $\tilde{e}_{n-i}$ and $\tilde{\eta}_i$ in the same way, using (4.3.15) and rearranging the variables and equations in (4.3.14), we can write

$$\begin{bmatrix} T_1 & h^2 T_2 \\ h^2 T_3 & hT_4 \end{bmatrix} \begin{bmatrix} \tilde{E}'^{(1)} \\ \tilde{E}'^{(2)} \end{bmatrix} = -\begin{bmatrix} S_1 & hS_2 \\ hS_3 & S_4 \end{bmatrix} \begin{bmatrix} \tilde{e}^{(1)}_{n-1} + \tilde{\delta}^{(1)} \\ \tilde{e}^{(2)}_{n-1} + \tilde{\delta}^{(2)} \end{bmatrix} + \begin{bmatrix} \tilde{\eta}^{(1)} \\ \tilde{\eta}^{(2)} \end{bmatrix} \quad (4.3.16)$$

where

$$\tilde{E}'^{(i)} = [\tilde{E}'^{(i)}_1, \tilde{E}'^{(i)}_2, \ldots, \tilde{E}'^{(i)}_M]^T,$$
$$\tilde{\delta}^{(i)} = [\tilde{\delta}^{(i)}_1, \tilde{\delta}^{(i)}_2, \ldots, \tilde{\delta}^{(i)}_M]^T,$$
$$\tilde{\eta}^{(i)} = [\tilde{\eta}^{(i)}_1, \tilde{\eta}^{(i)}_2, \ldots, \tilde{\eta}^{(i)}_M]^T,$$
$$\tilde{e}^{(i)}_{n-1} = [\tilde{e}^{(i)}_{n-1}, \tilde{e}^{(i)}_{n-1}, \ldots, \tilde{e}^{(i)}_{n-1}]^T$$

for $i = 1, 2$, and

$$T_1 = \hat{T}_1 + O(h),$$
$$T_4 = \hat{T}_4 + O(h),$$
$$S_1 = \hat{S}_1 + O(h),$$
$$S_4 = I + O(h)$$

where $\hat{T}_1 = I + h\mathcal{A} \otimes C$, $\hat{T}_4 = \mathcal{A} \otimes I$, $\hat{S}_1 = I \otimes C$, and T_2, T_3, S_2, S_3 are matrices whose elements are $O(1)$.

Let $\overline{T}_n$ denote the left-hand matrix in (4.3.16). $\overline{T}_n$ can be written as

$$\overline{T}_n = \begin{bmatrix} I & 0 \\ 0 & hI \end{bmatrix} \begin{bmatrix} \hat{T}_1 + O(h) & h^2 T_2 \\ hT_3 & \hat{T}_4 + O(h) \end{bmatrix}. \quad (4.3.17)$$

$\hat{T}_4$ is invertible because the matrix $\mathcal{A}$ of coefficients of the Runge-Kutta method is invertible. By inverting the right hand side of (4.3.17) the inverse of $\overline{T}_n$ is given by

$$\overline{T}_n^{-1} = \begin{bmatrix} \hat{T}_1^{-1} + O(h) & O(h) \\ O(h) & \frac{\hat{T}_4^{-1}}{h} + O(1) \end{bmatrix}. \quad (4.3.18)$$

4.3. NONLINEAR INDEX ONE SYSTEMS

Using (4.3.18) to solve (4.3.16) for $\tilde{E}'^{(1)}$ and $\tilde{E}'^{(2)}$, we have

$$\begin{bmatrix} \tilde{E}'^{(1)} \\ \tilde{E}'^{(2)} \end{bmatrix} = - \begin{bmatrix} \hat{T}_1^{-1}\hat{S}_1 + O(h) & O(h) \\ O(1) & \frac{\hat{T}_4^{-1}}{h} + O(1) \end{bmatrix} \times \qquad (4.3.19)$$

$$\begin{bmatrix} \tilde{e}_{n-1}^{(1)} + \tilde{\delta}^{(1)} \\ \tilde{e}_{n-1}^{(2)} + \tilde{\delta}^{(2)} \end{bmatrix} + \overline{T}_n^{-1} \begin{bmatrix} \tilde{\eta}^{(1)} \\ \tilde{\eta}^{(2)} \end{bmatrix}.$$

Multiplying (4.3.12b) by Q^{-1}, which we now denote by Q_n^{-1} to show its dependence upon $(t_n, y(t_n), y'(t_n))$ we obtain

$$Q_n^{-1} e_n = \tilde{e}_{n-1} + h \sum_{i=1}^{M} b_i \tilde{E}_i' + \tilde{\delta}_{M+1}. \qquad (4.3.20)$$

Inserting (4.3.19) into (4.3.20), we have

$$Q_n^{-1} e_n = S_n Q_n^{-1} e_{n-1} - h U_n \tilde{\delta}^{(n)} + \tilde{\delta}_{M+1}^{(n)} + h B \overline{T}_n^{-1} \tilde{\eta}^{(n)}, \qquad (4.3.21)$$

where

$$S_n = \left(I - h \begin{bmatrix} b_1^T \hat{T}_1^{-1} \hat{S}_1 + O(h) & O(h) \\ O(1) & \frac{b_2^T \hat{T}_4^{-1}}{h} + O(1) \end{bmatrix} \begin{bmatrix} Z_1 & 0 \\ 0 & Z_2 \end{bmatrix} \right),$$

$$U_n = \begin{bmatrix} b_1^T \hat{T}_1^{-1} \hat{S}_1 + O(h) & O(h) \\ O(1) & \frac{b_2^T \hat{T}_4^{-1}}{h} + O(1) \end{bmatrix},$$

$$Z_1 = \epsilon_M \otimes I_{m_1},$$
$$Z_2 = \epsilon_M \otimes I_{m_2},$$
$$b_1^T = b^T \otimes I_{m_1},$$
$$b_2^T = b^T \otimes I_{m_2},$$
$$B = \begin{bmatrix} b_1^T & 0 \\ 0 & b_2^T \end{bmatrix},$$
$$\tilde{\delta}^{(n)} = \begin{bmatrix} \tilde{\delta}^{(1)} \\ \tilde{\delta}^{(2)} \end{bmatrix}, \quad \tilde{\eta}^{(n)} = \begin{bmatrix} \tilde{\eta}^{(1)} \\ \tilde{\eta}^{(2)} \end{bmatrix}$$

where again $\epsilon_M = [1, 1, \cdots, 1]^T$. By the definition of $\hat{T}_4$, we have

$$b_2^T \hat{T}_4^{-1} Z_2 = (1-r) I_{m_2},$$

where r is the stability constant, $|r| \leq 1$. Thus S_n has the form

$$S_n = K + O(h), \qquad (4.3.22)$$

where
$$K = \begin{bmatrix} I_{m_1} & 0 \\ 0 & rI_{m_2} \end{bmatrix},$$
as in Lemma 4.3.1.

Solving for e_n in (4.3.21), we obtain

$$e_n = \prod_{j=0}^{n-1} Q_{n-j} S_{n-j} Q_{n-j}^{-1} e_0 + \qquad (4.3.23)$$

$$\sum_{i=0}^{n-1} \left[\left(\prod_{j=0}^{i-1} Q_{n-j} S_{n-j} Q_{n-j}^{-1} \right) Q_{n-i} \left(\tilde{\delta}_{M+1}^{(n-i)} - hU_{n-i}\tilde{\delta}^{(n-i)} + h\mathcal{B}\overline{T}_{n-i}^{-1} \tilde{\eta}^{(n-i)} \right) \right]$$

where $\prod_{j=0}^{i-1} Q_{n-j} S_{n-j} Q_{n-j}^{-1}$ is defined to be the identity for $i = 0$. Now,

$$\prod_{j=0}^{n-1} Q_{n-j} S_{n-j} Q_{n-j}^{-1} = Q_n \left(\prod_{j=0}^{n-2} S_{n-j} Q_{n-j}^{-1} Q_{n-j-1} \right) S_1 Q_1^{-1}$$

$$= Q_n (K + O(h))^n Q_1^{-1}$$

and

$$\prod_{j=0}^{i-1} Q_{n-j} S_{n-j} Q_{n-j}^{-1} = Q_n (K + O(h))^i Q_{n-i+1}^{-1}.$$

We can rewrite (4.3.23),

$$Q_n^{-1} e_n = (K + O(h))^n (I + O(h)) \tilde{e}_0 +$$
$$\qquad (4.3.24)$$
$$\sum_{i=0}^{n-1} (K + O(h))^i (I + O(h))(\tilde{\delta}_{M+1}^{(n-i)} - hU_{n-j}\tilde{\delta}^{(n-i)} + h\mathcal{B}\overline{T}_{n-i}^{-1} \tilde{\eta}^{(n-i)}).$$

Let

$$\hat{U}_n = \begin{bmatrix} b_1^T \hat{T}_1^{-1} \hat{S}_1 & 0 \\ 0 & \dfrac{b_2^T \hat{T}_4^{-1}}{h} \end{bmatrix}.$$

Rewriting (4.3.24), we obtain

$$\begin{aligned} \tilde{e}_n &= (K + O(h))^n (I + O(h)) \tilde{e}_0 \\ &+ \sum_{i=0}^{n-1} (K + O(h))^i (I + O(h))(\tilde{\delta}_{M+1}^{(n-i)} - h\hat{U}_{n-i}\tilde{\delta}^{(n-i)}) \\ &+ \sum_{i=0}^{n-1} (K + O(h))^i (I + O(h)) h\mathcal{B}\overline{T}_{n-i}^{-1} \tilde{\eta}^{(n-i)} \qquad (4.3.25) \\ &+ \sum_{i=0}^{n-1} (K + O(h))^i (I + O(h)) h(U_{n-i} - \hat{U}_{n-i}) \tilde{\delta}^{(n-i)}. \end{aligned}$$

4.3. NONLINEAR INDEX ONE SYSTEMS

Let k_d be the order of the IRK method applied to (nonstiff) ODE systems. Because $Q(t, y(t), y'(t))$ is a continuous function of t for uniform index one systems, and assuming the solution $y(t)$ to the DAE is sufficiently smooth, so that $\tilde{\delta}^{(n-i)}$ and $\tilde{\delta}^{(n-i)}_{M+1}$ are continuous, then $\tilde{\delta}^{(n+1)}_{M+1} - h\hat{U}_{n-i}\,\tilde{\delta}^{(n-i)}$ is also a continuous function of t. Similarly, we will assume $B\bar{T}^{-1}_{n-i}\tilde{\eta}^{(n-i)}$ is a continuous function of t, which is true under mild assumptions on the smoothness of y if the Newton errors are neglected. Finally, we know $\|\tilde{\delta}^{(n-i)}\| = O(h^{k_I+1})$. Using the facts $\delta^{(n-i)}_{M+1} = O(h^{k_d+1})$ and

$$b^T \mathcal{A}^{-1}\delta^{(n-i)} - \delta^{(n-i)}_{M+1} = O(h^{k_{a,1}+1}),$$

note that $\tilde{\delta}^{(n-i)}_{M+1} - h\hat{U}_{n-i}\,\tilde{\delta}^{(n-i)}$ is $O(h^{\min(k_d+1,k_I+2)})$ in the differential part and $O(h^{k_{a,1}+1})$ in the algebraic part. From Lemma 4.3.1, we have for $r \neq 1$

$$(K + O(h))^i = \begin{bmatrix} O(1) & O(h) \\ O(h) & |r|^i O(1) + O(h) \end{bmatrix}.$$

Partition (4.3.25) into differential and algebraic parts. Let $\bar{\eta}_1, \bar{\eta}_2$ be the Newton errors and suppose that the nonlinear higher order terms satisfy $\|\tilde{\eta}^{(i)}\| \leq \epsilon_i$ (this does not include the Newton errors) and the errors in the initial conditions satisfy $\|\tilde{e}^{(i)}_0\| = O(\zeta_i)$, $i = 1, 2$. Then we have from (4.3.25) the following cases:

(i) $-1 < r < 1$:

$$\begin{aligned}
\tilde{e}^{(1)}_n &= O(\zeta_1) + O(h\zeta_2) + O(h^{k_d}) + O(h^{k_{a,1}+1}) + O(h^{k_I+1}) \\
&\quad + O(\epsilon_1) + O(\epsilon_2) + O(\bar{\eta}_1) + O(\bar{\eta}_2) \quad (4.3.26a) \\
\tilde{e}^{(2)}_n &= O(h\zeta_1) + O(h\zeta_2) + O(h^{k_d+1}) + O(h^{k_{a,1}+1}) + O(h^{k_I+2}) \\
&\quad + O(h\epsilon_1) + O(\epsilon_2) + O(h\bar{\eta}_1) + O(\bar{\eta}_2) \quad (4.3.26b)
\end{aligned}$$

Hence
$$e_n = O(\zeta) + O(\bar{\eta}) + O(\epsilon) + O(h^G) \quad (4.3.27)$$

where
$$G = \min(k_d, k_I + 1).$$

(ii) $r = -1$:

$$\begin{aligned}
\tilde{e}^{(1)}_n &= O(\zeta_1) + O(h\zeta_2) + O(h^{k_d}) + O(h^{k_{a,1}+1}) + O(h^{k_I+1}) \\
&\quad + O(\epsilon_1) + O(\epsilon_2) + O(\bar{\eta}_1) + O(\bar{\eta}_2) \\
\tilde{e}^{(2)}_n &= O(h\zeta_1) + O(\zeta_2) + O(h^{k_d+1}) + O(h^{k_{a,1}+1}) + O(h^{k_I+2}) \\
&\quad + O(h\epsilon_1) + O(\epsilon_2) + O(h\bar{\eta}_1) + O(\tfrac{\bar{\eta}_2}{h})
\end{aligned}$$

Hence
$$e_n = O(\zeta) + O(\epsilon) + O\left(\frac{\bar{\eta}}{h}\right) + O(h^G) \quad (4.3.28)$$

where
$$G = \min(k_d, k_I + 1).$$

The better than expected results for $\tilde{e}_n^{(2)}$ in this case are due to cancellations in the algebraic part in the sums of (4.3.25) which come about because of alternating signs in the bottom right block of K when $r = -1$, coupled with the assumptions about smoothness. This can be easily seen by grouping the terms in the sums of (4.3.25) together two at a time, and then bounding the resulting sums. We also note that when $|r| = 1$ the Newton iterations have to be solved to $O(h^{G+1})$. If $r = 1$, then $(K + O(h))^i = O(1)$, $i \leq n$.

(iii) $r = 1$:

$$\begin{aligned}
\tilde{e}_n^{(1)} &= O(\zeta_1) + O(\zeta_2) + O(h^{k_d}) + O(h^{k_{a,1}}) + O(h^{k_I+1}) \\
&\quad + O(\epsilon_1) + O(\epsilon_2) + O(\bar{\eta}_1) + O(\bar{\eta}_2) \\
\tilde{e}_n^{(2)} &= O(\zeta_1) + O(\zeta_2) + O(h^{k_{a,1}}) + O(h^{k_I+1}) + O(h^{k_d+1}) \\
&\quad + O(\epsilon_1) + O(\epsilon_2) + O(\bar{\eta}_1) + O(\tfrac{\bar{\eta}_2}{h})
\end{aligned}$$

Hence
$$e_n = O(\zeta) + O\left(\frac{\bar{\eta}}{h}\right) + O(\epsilon) + O(h^G) \tag{4.3.29}$$

where
$$G = \begin{cases} \min(k_d, k_I), & \text{if } k_{a,1} = k_I \\ \min(k_d, k_I + 1), & \text{if } k_{a,1} \geq k_I + 1. \end{cases}$$

The better than expected terms $O(\epsilon_1)$, $O(\epsilon_2)$ in $\tilde{e}_n^{(2)}$ come about because $b^T A^{-1} \epsilon_M = 1 - r = 0$ in this case, so that the smoothness of $\tilde{\eta}$ (excluding Newton errors) implies the $O(h^{-1}\tilde{\eta}^{(2)})$ term in $B\bar{T}_{n-i}^{-1}\tilde{\eta}^{(n-i)}$ vanishes. Thus for linear problems ($\epsilon = 0$) we obtain the results in the statement of the theorem by noting that
$$\begin{aligned} C(q), B(q) &\Rightarrow k_I = q \\ C(q), B(q+1) &\Rightarrow k_d \geq q + 1. \end{aligned}$$

For nonlinear systems, we sketch the proof. Recall from Section 3.2 that without loss of generality we can assume F has the form
$$F(t, y, y') = \begin{bmatrix} F_1(t, y, y') \\ G(t, y) \end{bmatrix}.$$

Then $\eta_n^{(1)}$ consists of higher order terms of the form
$$\left(e_{n-1} + h\sum_{j=1}^{M} a_{ij} E_j' + \epsilon_i\right)^T F_{yy} \left(e_{n-1} + h\sum_{j=1}^{M} a_{kj} E_j' + \epsilon_k\right),$$

$$\left(e_{n-1} + h\sum_{j=1}^{M} a_{ij} E_j' + \epsilon_i\right)^T F_{yy'} E_k', \qquad E_i' F_{y'y'} E_k',$$

4.3. NONLINEAR INDEX ONE SYSTEMS

and $\eta_n^{(2)}$ consists of terms of the form

$$\left(e_{n-1} + h\sum_{j=1}^{M} a_{ij}E'_j + \epsilon_i\right)^T F_{yy} \left(e_{n-1} + h\sum_{j=1}^{M} a_{kj}E'_j + \epsilon_k\right).$$

Because P_n has the form given in Lemma 3.2.1, and $\tilde{\eta}_n = P_n\eta_n$, it follows that $\tilde{\eta}_n^{(1)}$ and $\tilde{\eta}_n^{(2)}$ are composed of terms of the same form as $\eta_n^{(1)}$ and $\eta_n^{(2)}$, respectively.

Now from (4.3.27), (4.3.28) and (4.3.29), we have that

$$e_n = O(h^G) + O(\epsilon_1) + O(\epsilon_2), \tag{4.3.30}$$

and it follows from (4.3.19) together with (4.3.30) that

$$E'_i = O(h^{G-1}) + O(\epsilon_1) + O\left(\frac{\epsilon_2}{h}\right). \tag{4.3.31}$$

According to the form of $\tilde{\eta}_n^{(1)}$ and $\tilde{\eta}_n^{(2)}$, and substituting from (4.3.30) and (4.3.31), we have

$$\|\tilde{\eta}_n^{(1)}\| \leq K_1\left(h^{G-1} + \epsilon_1 + \frac{\epsilon_2}{h}\right)^2$$
$$\|\tilde{\eta}_n^{(2)}\| \leq K_2(h^G + \epsilon_1 + \epsilon_2)^2.$$

Now, determine ϵ_1 and ϵ_2 as solutions of

$$\epsilon_1 = K_3\left(h^{2(G-1)} + \epsilon_1^2 + \frac{\epsilon_2^2}{h^2} + \frac{\epsilon_1\epsilon_2}{h} + h^{G-1}\epsilon_1 + h^{G-2}\epsilon_2\right)$$
$$\epsilon_2 = K_4(h^{2G} + \epsilon_1^2 + \epsilon_2^2 + \epsilon_1\epsilon_2 + h^G\epsilon_1 + h^G\epsilon_2) \tag{4.3.32}$$

by solving (4.3.32) by functional iteration $\underline{\epsilon} = G(\underline{\epsilon})$ with initial value $\underline{\epsilon}^{(0)} = O(h^{2(G-1)})$. For $G \geq 3$, we can use the contraction mapping theorem to conclude $\underline{\epsilon} = O(h^{2(G-1)})$, and we are done. For $G = 2$, we cannot apply the theorem directly, because $\|\partial G/\partial \underline{\epsilon}\| = O(1)$. But if we scale the variables by $\bar{\epsilon}_1 = \epsilon_1, \bar{\epsilon}_2 = \epsilon_2/\sqrt{h}$, then we can apply the strategy above. Finally, the result now follows from (4.3.30). For $G = 1$ it is not possible to use this argument to reach the conclusion. However, the result is proved in Chapter 3 for the most useful first order IRK method which is the implicit Euler method. □

It is important to note that Theorem 4.3.1 gives only a lower bound on the order of an IRK method applied to nonlinear index one DAE systems. Thus it is possible that there exist IRK methods which can achieve a higher order of accuracy on this class of problems than what is predicted by the theorem. With this in mind, we present the results of some numerical experiments which confirm that the order reduction predicted in the theorem does occur, and that a few methods achieve a higher order than the lower bounds predict.

The test problem was constructed to illustrate the effects of coupling between the differential and algebraic parts of the system. The problem is given by

$$\begin{bmatrix} 1 & -t \\ 0 & 0 \end{bmatrix} \begin{bmatrix} x' \\ y' \end{bmatrix} + \begin{bmatrix} 1 & -(1+t) \\ 0 & 1 \end{bmatrix} \begin{bmatrix} x \\ y \end{bmatrix} = \begin{bmatrix} 0 \\ \sin t \end{bmatrix} \quad (4.3.33)$$

with initial values given by

$$\begin{bmatrix} x(0) \\ y(0) \end{bmatrix} = \begin{bmatrix} 1 \\ 0 \end{bmatrix}, \quad \begin{bmatrix} x'(0) \\ y'(0) \end{bmatrix} = \begin{bmatrix} -1 \\ 1 \end{bmatrix}.$$

This problem has the solution

$$\begin{bmatrix} x(t) \\ y(t) \end{bmatrix} = \begin{bmatrix} \exp(-t) + t\sin t \\ \sin t \end{bmatrix}. \quad (4.3.34)$$

The problem was obtained from the constant coefficient index one DAE

$$\begin{aligned} \tilde{x}' &= -\tilde{x} \\ \tilde{y} &= \sin t \end{aligned} \quad (4.3.35)$$

by introducing a change of variables

$$\begin{bmatrix} x \\ y \end{bmatrix} = \begin{bmatrix} 1 & t \\ 0 & 1 \end{bmatrix} \begin{bmatrix} \tilde{x} \\ \tilde{y} \end{bmatrix}. \quad (4.3.36)$$

The test problem is uniform index one for all t, and because of the mixing introduced by the time-dependent transformation in (4.3.36), we would expect it to exhibit many of the order reduction effects described in Theorem 4.3.1.

In the experiments to determine the global error, we solved the problem with a sequence of fixed stepsizes over the interval [0,1]. The reported observed order of the global error reflects the behavior of the global error at the end of the interval as the stepsize is decreased by successive factors of two. To compute the observed local error, we solved (4.3.33) with various Runge-Kutta methods over one time step. The reported observed local error reflects the behavior of the error after one step as the stepsize is decreased by successive factors of two.

We experimented with several Runge-Kutta methods which might appear to be likely candidates for solving stiff or differential-algebraic systems. The methods were:

1. 2-stage, '2nd order' singly-implicit method (SIRK) [39], with $\lambda = 1 - \sqrt{2}/2$

2. 5-stage, '4th order' diagonally-implicit method (DIRK) [2,67]

3. 3-stage, '3rd order' singly-implicit method (SIRK) [39], with $1/\lambda$ the root of the Laguerre polynomial of degree three

4.3. NONLINEAR INDEX ONE SYSTEMS

4. 7-stage, '3rd order' extrapolation method based on implicit Euler method and polynomial extrapolation.

5. 3-stage '2nd order' L-stable semi-implicit method (Houbak and Thomsen [145])

6. 3-stage '4th order' Lobatto IIIC method (Chipman[68])

7. Implicit midpoint method

8. 2-stage A-stable Gauss-Legendre method ([138])

Table 4.3.1 gives the results of the experiments. In Table 4.3.1, k_d is the nonstiff order, $k_{a,1}$ is the order for constant-coefficient index one systems, k_I is the internal stage order, k_G is the lower bound predicted by Theorem 4.3.1, and k_g is the order of the observed global error.

Method	r	k_d	$k_{a,1}$	k_I	k_G	k_g
1. Two-stage SIRK	0	2	∞	2	2	2
2. Five-stage DIRK	0	4	∞	1	2	2
3. Three-stage SIRK	0	3	3	2	3	3
4. Seven-stage Extrp.	0	3	∞	1	2	3
5. Houbak & Thomsen	0	2	1	1	2	2
6. Lobatto IIIC	0	4	∞	2	3	4
7. Midpoint	-1	2	1	1	2	2
8. Gauss-Legendre	1	4	2	2	2	2

Table 4.3.1: Order for Nonlinear Index One Systems

Based on the results of Table 4.3.1, we can make a few observations. As expected, in no case was the lower bound for the order predicted by the theory higher than the order which was actually observed, and in many cases these two orders coincided. The observed orders for the extrapolation method and for the Lobatto IIIC formula were higher than would be expected based on the theorem.

Since k_I for a semi-explicit Runge-Kutta method is limited to one (because the first stage is necessarily an implicit Euler step), we would expect the order of the global errors for these methods to be limited to two. In particular, the DIRK methods are apparently limited to order two in most cases. Orders higher than two can be achieved by going to a fully-implicit formula such as the Lobatto IIIC method where the stage orders are higher. SIRK methods appear to be promising for DAE's in the light of these results, because high stage orders can be achieved and the methods can be implemented efficiently.

Necessary and sufficient order conditions for the local truncation error for IRK methods satisfying the stability condition applied to *linearly implicit* (i.e., DAE's in which the derivatives appear only linearly) and general index one DAE systems have recently been obtained by Kværnø [157,158]. The analysis makes use of the theory of Butcher series and rooted trees, as in [214]. Global order results based on a similar analysis are given by Hairer, Lubich and Roche [136] for linearly implicit DAE's. The main result improves the order of accuracy for some stiffly accurate formulas. In particular, the Radau IIA and Lobatto IIIC methods retain their nonstiff order of accuracy.

Collocation methods are an important subset of implicit Runge-Kutta methods with special properties. In particular, an M-stage collocation method always satisfies, by construction, $B(M)$ and $C(M)$. Thus the results in this section yield a lower bound on the global order for M-stage collocation methods applied to fully-implicit index one DAE systems of $O(h^M)$ for $-1 \leq r \leq 1$, and $O(h^{M+1})$ for $-1 \leq r < 1$. The order result of $O(h^{M+1})$ requires the additional assumption that the nonstiff order is at least $M+1$, and, for $r = -1$, that the stepsize is constant or changing very smoothly. Ascher [6] has recently given these results for Gaussian collocation methods, which satisfy $|r| = 1$.

We *do not recommend $|r| = 1$ schemes for the solution of fully-implicit index one initial value problems.* As we have noted in the previous section, implicit Runge-Kutta methods with $|r| = 1$ are in some sense unstable with respect to the accumulation of roundoff error. A more critical stability consideration, for $|r| = 1$ schemes applied to fully-implicit index one DAE's, has been pointed out by Ascher [7]. The stability of the scheme is controlled by the stability of an underlying ODE problem which is not necessarily stable when the original DAE system is stable. The asymptotic order results in this section are technically correct for $|r| = 1$ methods, but the order constant can be very large. The problem is most severe when the differential and algebraic parts of the system are tightly coupled together. Ascher illustrates with a test problem, which is a generalization of the problem given in this section, how the implicit midpoint method can yield global errors which grow exponentially in time, even though the underlying ODE system is stable and the errors are $O(h^2)$. When these schemes are applied to boundary value problems, this conditioning problem can sometimes be corrected by locating some of the consistency conditions corresponding to algebraic constraints at the correct boundary [7]. However, for initial value problems this is not practical because it transforms the initial value problem into a boundary value problem, which is more expensive to solve. The asymptotic order results for $|r| = 1$ schemes applied to DAE initial value problems are useful because they imply the order results for DAE boundary value problems. (In [77], it is shown that an implicit Runge-Kutta method is globally $O(h^k)$ for a properly formulated linear DAE boundary value problem iff it is $O(h^k)$ for the DAE formulated as a related initial value problem.) We do not have enough experience with $|r| = 1$ schemes applied to fully-implicit index one systems from applications to make a judgement on how often the type of problems that cause difficulty tend to occur.

4.4 Semi-Explicit Nonlinear Index Two Systems

In this section we study the behavior of IRK methods applied to semi-explicit nonlinear index two systems. The main result will be a convergence theorem which gives a set of order conditions which are sufficient to ensure that a method attains a given order for these systems. We also present some numerical experiments which illustrate the order reduction effects predicted by the theory.

Consider the semi-explicit nonlinear index two system

$$f(t, x, x', y) = 0 \quad (4.4.1a)$$
$$g(t, x, y) = 0, \quad (4.4.1b)$$

where we assume that $(\partial f/\partial x')^{-1}$ exists and is bounded in a neighborhood of the solution, $\partial g/\partial y$ has constant rank, and f and g have as many continuous partial derivatives as desired in a neighborhood of the solution. Defining

$$A_2(q): \begin{array}{l} b^T A^{-1} \epsilon_M = b^T A^{-2} c \\ b^T A^{-2} c^i = i, \quad i = 2, 3 \ldots, q \end{array}$$

(note that A_2 are the order conditions for linear constant coefficient index two systems), the principle result we prove in this section is

Theorem 4.4.1 *Given the nonlinear semi-explicit index two system (4.4.1) to be solved numerically by the M-stage IRK method (4.1.2), suppose that*

1. *The IRK method satisfies the strict stability condition $|r| < 1$*
2. *The errors in the initial conditions satisfy $\|e_0\| = O(h^{G_x})$, where G_x is defined below*
3. *The errors in terminating the Newton iterations are $O(h^{G_x+1})$*
4. *$G_x > 1$.*

Then the global errors in x_n and y_n are $O(h^{G_x})$ and $O(h^{G_y})$ respectively, where G_x and G_y are given by

$$G_x = \begin{cases} q, & \text{if } C(q), B(q) \\ q+1, & \text{if } C(q), B(q+1) \end{cases}$$

$$G_y = \begin{cases} q, & \text{if } C(q), B(q), A_1(q) \\ q+1, & \text{if } C(q), B(q+1), A_1(q+1), A_2(q+1). \end{cases}$$

Proof. Consider the semi-explicit nonlinear index two system (4.4.1). As noted by Gear [118], if we let $w' = y$, then the related nonlinear system

$$\begin{array}{l} f(t, x, x', w') = 0 \\ g(t, x, w') = 0 \end{array} \quad (4.4.2)$$

is index one. It is easy to see that solving the index one system (4.4.2) by the IRK method gives exactly the same solution for x as solving the original semi-explicit index two system (4.4.1) by the IRK method. Thus the error in x is given by Theorem 4.3.1.

Now consider the error in y. The numerical solution for y satisfies

$$\begin{aligned} Y^{(n)} &= y_{n-1} \otimes \epsilon_M + h(\mathcal{A} \otimes I)Y'^{(n)} \\ y_n &= y_{n-1} + hb_M^T Y'^{(n)} \end{aligned} \quad (4.4.3)$$

where $Y^{(n)} = [Y_1^{(n)}, Y_2^{(n)}, \ldots, Y_M^{(n)}]^T$ is a vector of stage approximations to y, $Y'^{(n)} = [Y_1'^{(n)}, Y_2'^{(n)}, \ldots, Y_M'^{(n)}]^T$ is a vector of stage approximations to the derivatives of y, $y_{n-1} \otimes \epsilon_M = [y_{n-1}^T, y_{n-1}^T, \ldots, y_{n-1}^T]^T$, and $b_M^T = b^T \otimes \epsilon_M$. Subtracting from (4.4.3) the corresponding expressions involving the true solution, we obtain

$$\begin{aligned} E^{y(n)} &= e_{n-1}^y \otimes \epsilon_M + h(\mathcal{A} \otimes I)E^{y'(n)} + \delta^{y(n)} & (4.4.4a) \\ e_n^y &= e_{n-1}^y + hb_M^T E^{y'(n)} + \delta_{M+1}^{y(n)}, & (4.4.4b) \end{aligned}$$

where $E^{y(n)} = Y^{(n)} - Y^{(n)}(t)$, $E^{y'(n)} = Y'^{(n)} - Y'^{(n)}(t)$, and $\delta^{y(n)}$, $\delta_{M+1}^{y(n)}$ are the internal local truncation errors for the y variable. Solving for $E^{y'(n)}$ in (4.4.4a) and substituting into (4.4.4b), we have

$$e_n^y = e_{n-1}^y + b_M^T(\mathcal{A}^{-1} \otimes I)\left(E^{y(n)} - e_{n-1}^y \otimes \epsilon_M - \delta^{y(n)}\right) + \delta_{M+1}^{y(n)}. \quad (4.4.5)$$

Rewriting (4.4.5) and noting that $b_M^T(\mathcal{A}^{-1} \otimes I)\delta^{y(n)} - \delta_{M+1}^{y(n)} = O(h^{k_{a,1}+1})$ and that $1 - b^T \mathcal{A}^{-1}\epsilon_M = r$, we have

$$e_n^y = re_{n-1}^y + b_M^T(\mathcal{A}^{-1} \otimes I)E^{y(n)} + O(h^{k_{a,1}+1}). \quad (4.4.6)$$

Substituting $E^{y(n)} = E^{w'(n)}$ in (4.4.6), we obtain

$$e_n^y = re_{n-1}^y + b_M^T(\mathcal{A}^{-1} \otimes I)E^{w'(n)} + O(h^{k_{a,1}+1}). \quad (4.4.7)$$

Now note that w is a variable in the index one system, so an expression for $\tilde{E}^{w'(n)}$ is given by (4.3.23). Substituting (4.3.25) for $\tilde{e}_{n-1}$ in (4.3.19) (using the strict stability condition $|r| < 1$), we have

$$b_M^T(\mathcal{A}^{-1} \otimes I)\tilde{E}^{w'(n)} = \begin{pmatrix} 0 \\ \frac{1}{h}b_M^T(\mathcal{A}^{-2} \otimes I_{m_2})\tilde{\delta}^{(2)} + O(h^{k_{a,1}}) \end{pmatrix} + O(h^{G_x}).$$

Now since $\tilde{\delta}^{(2)} = \left[\left(\mathcal{A}c^{k_I} - \frac{c^{k_I+1}}{k_I+1}\right) \otimes I_{m_2}\right] O(h^{k_I+1})$, we have

$$\frac{1}{h}b_M^T(\mathcal{A}^{-2} \otimes I_{m_2})\tilde{\delta}_2 = \left(b^T \mathcal{A}^{-1}c^{k_I} - b^T \mathcal{A}^{-2}\frac{c^{k_I+1}}{k_I+1}\right) \otimes I_{m_2} O(h^{k_I}).$$

4.4. SEMI-EXPLICIT NONLINEAR INDEX TWO SYSTEMS

This term is zero if $k_{a,2} \geq k_I + 1$. Thus we can conclude that

$$E^{w'} = \begin{cases} O(h^{G_x}) + O(h^{k_{a,1}}), & \text{if } k_{a,2} \geq k_I + 1 \\ O(h^{G_x}) + O(h^{k_{a,1}}) + O(h^{k_I}), & \text{otherwise.} \end{cases} \quad (4.4.8)$$

We consider two cases. First let $C(q)$, $B(q)$, and $A_1(q)$ hold. Then $k_I = q$, $k_{a,1} \geq q$, and $G_x = q$. Summing the recurrence (4.4.7) and using the strict stability condition, we obtain

$$\begin{aligned} e_n^y &= O(h^{G_x}) + O(h^{k_{a,1}}) + O(h^{k_I}) \\ &= O(h^q). \end{aligned}$$

In the second case, let $C(q)$, $B(q+1)$, $A_1(q+1)$, and $A_2(q+1)$ hold. Then $k_I = q$, $k_d \geq q+1$, $G_x = q+1$, $k_{a,1} \geq q+1$, and $k_{a,2} \geq q+1$, so $k_{a,2} \geq k_I$. Again summing the recurrence (4.4.7), it follows that

$$\begin{aligned} e_n^y &= O(h^{G_x}) + O(h^{k_{a,1}}) \\ &= O(h^{q+1}). \square \end{aligned}$$

The reader may notice that order results for $|r| = 1$ methods applied to semi-explicit index two systems have not been given here. Because of their close relationship to fully-implicit index one systems, the index two systems can also yield conditioning difficulties for $|r| = 1$ schemes. Further, the order for the y variables is generally quite poor for these methods. While the order loss can sometimes be corrected by differencing the y variable with a different formula than the x variable, the stability problem, when it is present, is more severe [7]. Thus, these schemes are not recommended for semi-explicit index two DAE initial value problems.

As in the case of IRK methods applied to nonlinear index one systems, it is important to note that these results give only a lower bound on the order that the IRK method can achieve. We tested the L-stable IRK methods from Table 4.2.1 on two nonlinear problems. We chose to study the index three pendulum problem simply because it has been studied so frequently by DAE researchers [119,121,160] and can be posed as an index two problem [121]. The other nonlinear problem considered arises in the context of trajectory prescribed path control problems, a topic introduced earlier in Chapter 1. The exact solution is not available for either problem, so we first had to generate a 'true' solution which could be used for comparison. The corresponding index one systems were formulated and solved by the code DASSL [200] with extremely tight error tolerances.

We first tested the pendulum problem as formulated in [121]. This formulation ensures that the original index three algebraic constraint is satisfied even though the index of the system has been reduced to two. The system is given by

$$\begin{aligned}
x_1' &= x_3 - x_1 y_2 \\
x_2' &= x_4 - x_2 y_2 \\
x_3' &= -y_1 x_1 \\
x_4' &= -y_1 x_2 - 1 \\
0 &= (1 - x_1^2 - x_2^2)/2 \\
0 &= x_1 x_3 + x_2 x_4.
\end{aligned} \qquad (4.4.9)$$

The algebraic constraints in this problem are nonlinear, yet the algebraic variables appear only linearly in the system. The pendulum problem was solved using the fixed stepsize IRK code on the interval [0,1] for a sequence of stepsizes with each particular IRK formula. Consistent initial conditions were specified, namely $x_1 = 1$, $x_2 = x_3 = x_4 = y_1 = y_2 = 0$. The corrector iteration was terminated with a tolerance of 10^{-8}, because the Newton iteration failed to converge for tighter tolerances. Rates of convergence for each method were estimated as in the linear problem by comparing the global errors at $t = 1$ for numerical solutions produced by successively halving the stepsize. The numerical results for the pendulum problem are summarized in Table 4.4.1.

Method	k_d	$k_{a,1}$	$k_{a,2}$	k_I	k_G^x	k_g^x	k_G^y	k_g^y
1. Two-stage SIRK	2	∞	2	2	2	2	2	2
2. Five-stage DIRK	4	∞	1	1	2	2	1	1
3. Three-stage SIRK	3	3	2	2	3	3	2	2
4. Seven-stage Extrp.	3	∞	3	1	2	3	2	3

Table 4.4.1: Predicted and Observed Orders on Pendulum Problem

Next, we tested an index two trajectory problem which is representative of trajectory prescribed path control problems of current interest. In Section 6.3 this application will be addressed in more detail, and we defer the task of specifying the DAE system until then. Initial values for the state variables are known exactly, but initial values for the two algebraic variables (namely, angle of attack α and bank angle β) were determined numerically from the corresponding index one system. Specifically, the test problem used the following initial values for the state variables: altitude $H = 100,000$ feet, longitude $\xi = 0°$, latitude $\lambda = 0°$, relative velocity $V_R = 12000$ feet/second, flight path angle $\gamma = 0°$, and azimuth $A = 45°$. Angle of attack and bank angle were initialized to $\alpha = 2.672870042°$ and $\beta = -.0522095861634°$, respectively. The 'small' errors in the initial values for the algebraic variables were annihilated in one step by the IRK methods chosen, as a result of their L-stability property. The Newton iteration was terminated with a tolerance of 10^{-10}. This problem was solved for fixed stepsizes on the interval [0, 300], and the global errors in the solution were computed at $t = 300$ using the 'true' solution described earlier.

4.4. SEMI-EXPLICIT NONLINEAR INDEX TWO SYSTEMS

Since the corrector iteration was terminated with a fairly tight tolerance, the values of the two state variables prescribed by the algebraic constraints (namely, the flight path angle γ and the azimuth A) were computed almost exactly for all the IRK methods considered. The three-stage SIRK method surprised us by producing a third order accurate solution for the algebraic variables as well as for the state variables. We suspect that this difference in performance for this particular IRK method, when compared to its results on the linear test problem and the pendulum problem, must be due to the specific coupling of the state and algebraic variables in this nonlinear system. The numerical results for the trajectory problem are summarized in Table 4.4.2.

Method	k_d	$k_{a,1}$	$k_{a,2}$	k_I	k_G^x	k_q^x	k_G^y	k_q^y
1. Two-stage SIRK	2	∞	2	2	2	2	2	2
2. Five-stage DIRK	4	∞	1	1	2	2	1	1
3. Three-stage SIRK	3	3	2	2	3	3	2	2
4. Seven-stage Extrp.	3	∞	3	1	2	3	2	3

Table 4.4.2: Predicted and Observed Orders on Trajectory Problem

In conclusion, we see that the observed convergence rates of these IRK methods applied to nonlinear semi-explicit index two systems can sometimes be as slow as the lower bounds derived in Theorem 4.4.1 would indicate. Some formulas, in particular the extrapolation method, achieve an order of accuracy exceeding the predicted lower bounds.

There is a class of IRK methods which have an internal order as high or nearly as high as the ODE order. In particular, consider the class of M-stage singly implicit Runge-Kutta methods (SIRK's) whose coefficient matrix $\mathcal{A}$ is characterized by its single-fold eigenvalue. Butcher [45] has shown how these IRK formulas can be implemented very efficiently. There are two types of SIRK's, the *transformed* type [39] and the *collocation* type [194]. It is easy to show that for index two problems, transformed SIRK's will be at least order $M - 1$ (since $k_I \geq M - 1$), while collocation SIRK's will be order M (since $k_I = M$). If an L-stable SIRK formula is desired, one may select the eigenvalue of the $\mathcal{A}$ matrix to satisfy $L_M(\lambda^{-1}) = 0$ where L_M is the Laguerre polynomial of degree M. Methods of this type have been derived for orders up to and including six. Thus we expect the SIRK methods to perform very well on index two problems. However, the development of an efficient IRK code for DAE's with index greater than one remains a challenge because of the difficulties in developing appropriate error control strategies for all the variables.

We note that in a recent paper, Hairer, Lubich and Roche [136] have derived sharper order conditions for the x variable, for index two systems of Hessenberg form. Order results for the y variable remain the same as those given here.

Finally, we note that Hairer, Lubich and Roche [136] have obtained a preliminary convergence result for index three Hessenberg systems

$$\begin{aligned} x_1' &= F_1(x_1, x_2, x_3, t) \\ x_2' &= F_2(x_1, x_2, t) \\ 0 &= F_3(x_2, t), \end{aligned}$$

which yields estimates of the global error of $O(h^{k_I})$ for x_1 and x_2, and $O(h^{k_I-1})$ for x_3, where k_I is the stage order, $k_I \geq 2$, and $|r| < 1$.

4.5 Order Reduction and Stiffness

There is a close relationship between DAE systems and stiff ODE systems. For example, consider the singular perturbation problem

$$\begin{aligned} y' &= f(y, z, \epsilon) \\ \epsilon z' &= g(y, z, \epsilon), \quad 0 < \epsilon \ll 1 \end{aligned} \quad (4.5.1)$$

with initial values $y(0)$, $z(0)$ admitting a smooth solution (i.e., all derivatives of $y(t)$, $z(t)$ up to a sufficiently high order are bounded independently of ϵ.) The functions f and g are assumed to be smooth, and g must satisfy

$$\mu\left(\frac{\partial g}{\partial z}\right) \leq -1$$

in a neighborhood of the solution, where μ denotes the logarithmic norm with respect to some inner product. This problem tends to the semi-explicit index one DAE

$$\begin{aligned} y' &= f(y, z) \\ 0 &= g(y, z) \end{aligned} \quad (4.5.2)$$

as $\epsilon \to 0$, as do its solutions. Thus it is natural to expect that there will be a relationship between the behavior of numerical ODE methods when applied to the stiff problem, and their behavior when applied to the related DAE. In this section we will examine this relationship for implicit Runge-Kutta methods applied to index one DAE's and related stiff systems.

Order reduction (from the traditional nonstiff order) for Runge-Kutta methods applied to stiff ODE's was first noted in 1974 by Prothero and Robinson [206]. They studied the stiff model problem

$$y'(t) = \lambda y(t) + g'(t) - \lambda g(t), \quad \lambda \ll 0. \quad (4.5.3)$$

It is interesting to note that the order conditions for IRK methods applied to linear constant coefficient index one DAE's can also be derived by considering the test equation (4.5.3) and letting the stiffness $\lambda \to -\infty$ [42].

In 1981 a more comprehensive theory of order reduction for IRK methods applied to stiff ODE's was developed by Frank, Schneid and Ueberhuber [103,104]. In these important papers, the concepts of B-consistency and

4.5. ORDER REDUCTION AND STIFFNESS

B-convergence, which are the counterparts for stiff ODE's of the classical concepts of stability and convergence, were introduced. The monograph by Dekker and Verwer [93] provides an excellent introduction to the subject of B-convergence. Here we give only a brief overview to illustrate the relationship of this theory to order reduction theories for DAE's. Using the B-convergence theory, a lower bound for the order can be obtained which is valid over a class of stiff problems. In their initial work, Frank, Schneid and Ueberhuber studied the class of stiff ODE's $y' = f(t,y)$ satisfying, in addition to the usual assumptions for nonstiff ODE's, the one-sided Lipschitz condition

$$< f(t,y_1) - f(t,y_2), y_1 - y_2 > \leq \gamma \|y_1 - y_2\|^2, \quad \forall t \in \mathcal{R}, \forall y_1, y_2 \in \mathcal{R}^m, \quad (4.5.4)$$

and with initial values admitting a smooth solution. For this quite general class of stiff problems, they determined that the order q of an IRK method satisfies

$$k_I \leq q \leq k_d, \quad (4.5.5)$$

where k_I is the internal stage order as defined in Section 4.3.2, and k_d is the traditional (nonstiff) order of the method.

Subsequent research on order reduction for IRK methods applied to stiff ODE's has focused on obtaining sharper order bounds [105,106]. A larger lower bound for the order can sometimes be obtained by restricting the class of problems. Burrage, Hundsdorfer and Verwer [43], by considering the order of convergence of Runge-Kutta methods when applied to stiff semilinear systems of the form

$$y'(t) = Qy(t) + g(t, y(t)) \quad (4.5.6)$$

have shown that in many cases the global order associated with the B-convergence theory is one higher than the stage order. This analysis was extended to nonlinear problems satisfying the one-sided Lipschitz condition (4.5.4) by Burrage and Hundsdorfer [41]. By examining the global error recursion obtained by Burrage, Hundsdorfer and Verwer for (4.5.6) and letting the stiffness become infinite, it can be seen [42] that these order conditions tend precisely to the order conditions given in Section 4.3.2 for index one DAE's. Since some methods attain a higher order of B-convergence for the semilinear problem (4.5.6) than for the general nonlinear dissipative problem satisfying (4.5.4), the order of optimal B-convergence for general nonlinear dissipative problems may be one lower in some cases than the order for fully-implicit index one DAE's.

It is possible to derive DAE order results directly from the order results for stiff ODE's by constructing a stiff ODE whose solution tends to the solution of the original DAE, as in Knorrenschild [153]. This stiff ODE is called a *regularization* of the DAE. However, sharper order results for the DAE have so far been obtained by considering the DAE directly. It should be noted, though, that regularizations are important for other reasons, including the fact that many DAE systems from applications have their beginnings as stiff

ODE systems, where the stiffness has been taken to be infinite as a simplifying assumption for the physical problem.

It is also possible to obtain order results for certain classes of stiff ODE's from DAE order results. Hairer, Lubich and Roche [136] have obtained order results for the singular perturbation problem (4.5.1) in this way. The basic idea is to expand the solution in powers of the small parameter ϵ. The coefficients in the expansion of the global error in powers of ϵ are the global errors of the IRK methods applied to a differential-algebraic system which, for higher powers of ϵ, is a higher index DAE of a special form. The order behavior for the stiff problem (4.5.1) is in general somewhat worse than for the limiting DAE (4.5.2). Thus the stiff problem (4.5.1) is, in some sense, harder to solve than the infinitely stiff (DAE) system (4.5.2). Of course, the results for (4.5.1) tend in the limit of $\epsilon \to 0$ to results for (4.5.2). Numerical results in [136] illustrate the order reduction for some singular perturbation problems as a function of ϵ, and confirm the order reduction predicted by the theory. Finally, we note that these results implicitly assume that the stiff problem has been started on the smooth solution.

4.6 Extrapolation Methods

Although extrapolation methods may be viewed as IRK methods, it is natural to consider them in a class by themselves. In this section, we discuss the recent advances in theory and practice for extrapolation methods applied to DAE's. For a discussion on extrapolation methods for stiff and nonstiff ODE's, we refer the reader to the paper by Deuflhard [95].

As we have discussed earlier, one-step methods such as extrapolation have an inherent advantage over BDF methods when applied to DAE's with frequent discontinuities. There is a close connection between the behavior of extrapolation methods when applied to stiff ODE's and to DAE's. An early extrapolation code for stiff ODE's was developed by Dahlquist and Lindberg [87], and was based on the implicit trapezoid method. More recently, Bader and Deuflhard developed a stiff ODE code METAN1 [13,94], which is an implementation of an A-stable discretization based on the semi-implicit midpoint method. For semi-explicit index one DAE's, the code LIMEX [96], developed by Deuflhard et al., implements an extrapolation of the semi-implicit Euler method. This code has been used, for example, in the solution of problems arising in combustion modeling [177], where there are frequent discontinuities in time which arise because of an adaptive mesh strategy for the partial differential equations (see Section 6.5).

We begin by describing the extrapolation method based on the implicit Euler method when applied to a fully-implicit DAE,

$$F(t, y, y') = 0.$$

Let $y_h(t) = y_n$ denote the numerical solution at $t = nh$. Suppose that H is

4.6. EXTRAPOLATION METHODS

the basic stepsize, and that a sequence of positive integers $n_1 < n_2 < n_3 < \ldots$ is chosen which define the stepsizes $h_j = H/n_j$ such that $h_1 > h_2 > h_3 > \ldots$. Then an extrapolation method may be defined as follows

$$\begin{aligned} T_{j1} &= y_{h_j}(H) \\ T_{j,k+1} &= T_{j,k} + \frac{T_{j,k} - T_{j-1,k}}{(n_j/n_{j-k}) - 1} \end{aligned} \quad (4.6.1)$$

where $y_{h_j}(H)$ is the numerical solution at $t = H$ determined by taking n_j steps of length h_j of the implicit Euler method. Now if the discretization error $e(H; h)$ of the implicit Euler method can be written as an asymptotic expansion in the stepsize, namely

$$e(H; h) = \sum_{j=1}^{N} g_j(H) h^j + G_{N+1}(H; h) h^{N+1},$$

then the values in the kth column of the extrapolation tableau can be shown to be of order k. The proof of such an expansion includes the derivation of uniform bounds

$$\|G_{N+1}(t; h)\| \leq M_{N+1}(t) \quad \text{for } h \in [0, H].$$

For a simple nonstiff ODE $y' = f(t, y)$, and the choice of the integers n_j as the harmonic sequence $\{1, 2, 3, 4, 5, 6, \ldots\}$, one can show such an asymptotic error expansion does exist, and that

$$\begin{aligned} g_j(0) &= 0 \\ M_{N+1}(t) &= C_{N+1}(e^{Lt} - 1)/L, \quad L \neq 0 \end{aligned}$$

where L is the Lipschitz constant for f [95]. For a stiff ODE, the Lipschitz constant will be large, and hence a more useful characterization in terms of the logarithmic norm μ of f_y [87] and the deflated Lipschitz constant $\bar{L}$ [13] is preferred. However, as far as we know, a uniform bound on the remainder term involving μ and $\bar{L}$ has not been obtained [95]. One advantage to this specific extrapolation method based on the implicit Euler method is the fact that it is L-stable.

Consider a stiff system of the form

$$\begin{aligned} y' &= f(y, z), \quad y(0) = y_0 \\ \epsilon z' &= g(y, z), \quad z(0) = z_0, \ 0 < \epsilon \ll 1 \end{aligned} \quad (4.6.2)$$

and the corresponding semi-explicit index one DAE obtained by setting $\epsilon = 0$. It is assumed that the functions f and g are smooth and may depend on ϵ. In addition, we shall assume that a sufficiently high number of the derivatives of the solution $y(t)$, $z(t)$ to (4.6.2) can be bounded independently of ϵ, in which case the solution is said to be *smooth*. The existence of *perturbed* asymptotic expansions of the discretization error associated with general one-step methods of the form

$$y_{n+1} = y_n + h\phi(y_n, z_n, h)$$
$$z_{n+1} = \psi(y_n, z_n, h)$$

where the methods are only formally explicit, is shown in [96] in the limiting case (i.e. when $\epsilon = 0$). These one-step methods are assumed to be consistent to order p - i.e.,

$$y(t+h) - y(t) - h\phi(y(t), z(t), h) = O(h^{p+1})$$
$$z(t+h) - \psi(y(t), z(t), h) = O(h^p)$$

and to satisfy the contractivity condition

$$\left\| \frac{\partial \psi(y, z, 0)}{\partial z} \right\| \leq \alpha < 1$$

in a neighborhood of the solution. Then in general the global error corresponding to a one-step method has an expansion of the form ($t_n = nh \leq t$),

$$\begin{aligned}
y_n &= y(t_n) + h^p a_p(t_n) + h^{p+1}\left(a_{p+1}(t_n) + \beta_n^{p+1}\right) + \cdots \\
&\quad + h^N \left(a_N(t_n) + \beta_n^N\right) + h^{N+1} A(n, h) \\
z_n &= z(t_n) + h^p b_p(t_n) + h^{p+1}\left(b_{p+1}(t_n) + \gamma_n^{p+1}\right) + \cdots \\
&\quad + h^N \left(b_N(t_n) + \gamma_n^N\right) + h^{N+1} B(n, h)
\end{aligned}$$

where all coefficients vanish for $n = 0$ and $\beta_n^j(h)$, $\gamma_n^j(h) = O(\bar{\alpha}^n)$ for some $\bar{\alpha}$ with $\alpha < \bar{\alpha} < 1$. The remainder terms $A(n, h)$ and $B(n, h)$ are uniformly bounded for $nh \leq t$. Therefore, unperturbed asymptotic expansions of the global error do not exist in general for all one-step methods. However, we shall see that there are some particular methods on which extrapolation techniques may be based.

The existence of perturbed expansions is not limited to DAE's, but is in fact an inherent difficulty associated with applying extrapolation techniques to stiff ODE's of the form (4.6.2). To illustrate we state the following result, obtained by Hairer and Lubich [135]:

Theorem 4.6.1 *The numerical solution y_n, z_n satisfying the difference equations*

$$\begin{aligned}
y_n - y_{n-1} &= hf(y_n, z_n) \\
\epsilon(z_n - z_{n-1}) &= hg(y_n, z_n)
\end{aligned}$$

4.6. EXTRAPOLATION METHODS

obtained by discretizing (4.6.2) by the implicit Euler method, possesses for $h/\epsilon \geq 1$ a perturbed asymptotic expansion of the form

$$\begin{aligned}
y_n &= y(t_n) + ha_1(t_n) + h^2 a_2(t_n) + \ldots + h^N a_N(t_n) + O(h^{N+1}) \\
&\quad + h\epsilon^2 f_z(0) g_z^{-1}(0) \left(I - \frac{h}{\epsilon} g_z(0)\right)^{-n} \nu + O(h^2 \epsilon^2) \\
z_n &= z(t_n) + hb_1(t_n) + h^2 b_2(t_n) + \ldots + h^N b_N(t_n) + O(h^{N+1}) \\
&\quad + h\epsilon \left(I - \frac{h}{\epsilon} g_z(0)\right)^{-n} \nu + O(h^2 \epsilon)
\end{aligned} \quad (4.6.3)$$

where $t_n = nh$, $g_z(t) = g_z(y(t), z(t))$, $f_z(t) = f_z(y(t), z(t))$, the functions $a_i(t)$, $b_i(t)$ are smooth (i.e., their derivatives up to a certain order are bounded independently of ϵ), and

$$\begin{aligned}
a_i(0) &= O(\epsilon^2), \quad b_i(0) = O(\epsilon), \quad i = 1, 2, \ldots \\
\nu &= \frac{1}{2} g_z^{-2}(0) \Big(g_y(0) y''(0) + g_z(0) z''(0)\Big) + O(\epsilon).
\end{aligned}$$

The constants involved in the $O(\ldots)$ terms are independent of h, n, and ϵ.

For a complete statement of additional assumptions required by this theorem see [135]. Note that this theorem holds for system (4.6.2) when the logarithmic norm of $g_z(t)$ is negative on $[0, \bar{t}]$ for some $\bar{t}$.

It is particularly interesting that the pertubations in the expansion (4.6.3) vanish for the differential-algebraic system ($\epsilon = 0$). In this case, extrapolation may be applied in the usual way. However, when $\epsilon \neq 0$, the dominant terms in the asymptotic expansions for both the stiff and non-stiff components involve the expressions $h[I - (h/\epsilon)g_z(0)]^{-n}$. To understand how these pertubations affect the extrapolation tableau, Hairer and Lubich carefully analyze these terms with respect to the step number sequence $\{n_j\} = \{2, 3, 4, 6, 8, 12, 16, 24, 32, 48, \ldots\}$. The effect of these pertubations depends on whether H/ϵ is small (the nonstiff situation), large (the stiff situation), or of moderate size.

Much of the recent research on extrapolation methods has centered around the development of a *linearly implicit* method (also called *semi-implicit*) based on either the implicit Euler method or the implicit midpoint rule. A linearly implicit method requires only one linear system to be solved on each step, in contrast to a fully-implicit method where the Newton iteration is iterated until the errors are smaller than some tolerance. Deuflhard et al.[96] analyze a semi-implicit Euler discretization applied to semi-explicit index one DAE's, showing that the asymptotic error expansion of the global error does in fact contain perturbation terms. These results explain the pattern observed in the extrapolation tableau obtained for this method when applied with the stepsize sequence $h_i = H/i$, $i = 1, 2, \ldots, 10$. For convenience, we give the order patterns for extrapolation tableaus from [96] for y and z here in Tables 4.6.1 and 4.6.2.

```
1
1 2
1 2 3
1 2 3 4
1 2 3 4 4
1 2 3 4 4 5
1 2 3 4 4 5 5
1 2 3 4 4 5 6 5
1 2 3 4 4 5 6 6 5
1 2 3 4 4 5 6 7 6 5
```

Table 4.6.1: Order Pattern for Extrapolated y Variables

```
2
2 2
2 2 3
2 2 3 4
2 2 3 4 4
2 2 3 4 5 4
2 2 3 4 5 5 4
2 2 3 4 5 6 5 4
2 2 3 4 5 6 6 5 4
2 2 3 4 5 6 7 6 5 4
```

Table 4.6.2: Order Pattern for Extrapolated z Variables

The analysis of extrapolation methods for DAE's has recently been extended by Lubich [171] to more general index one DAE's in the form,

$$B(y)y' = f(y), \quad y(0) = y_0 \qquad (4.6.4)$$

on $[0, T]$. This work is a natural extension of analysis conducted earlier by Deuflhard, Hairer, and Zugck [96] where $B(y) = B$ is constant.

Several different formulations have been proposed for a linearly implicit method based on the implicit Euler method. Here we consider two such schemes defined for system (4.6.4) [171].

Scheme 1:

$$\left(B(y_0) + h\left(B'(y_0)y_0' - f'(y_0)\right)\right)(y_{n+1} - y_n) = hf(y_n) - \left(B(y_n) - B(y_0)\right)(y_n - y_{n-1}).$$

4.6. EXTRAPOLATION METHODS

Scheme 2:

$$\left(B(y_n) + h\left(B'(y_0)y_0' - f'(y_0)\right)\right)(y_{n+1} - y_n) = hf(y_n).$$

Scheme 2 was first proposed by Deuflhard and Nowak [97], although they did not investigate the existence of an asymptotic error expansion. Note that the iteration matrix in Scheme 2 changes on each step. Both schemes require y_0', which for a DAE is not immediately available (as $B(y)$ is singular). On the initial step, one might use the implicit Euler method with a very small time step to generate the y_0' values required. On subsequent steps, Deuflhard and Nowak propose to apply extrapolation to the sequence of values

$$z_h(H) = \frac{y_h(H) - y_h(H-h)}{h}$$

which approximate $y'(H)$.

For comparison to the result given earlier for the implicit Euler method, we state Lubich's results [171] concerning the existence of perturbed asymptotic error expansions for these two schemes.

Theorem 4.6.2 *For the index one system* (4.6.4), *the numerical solution determined by Scheme 1 has the following expansion for* $n^2h \leq K$, K *a sufficiently small constant and* $t_n = nh$

$$y_n = y(t_n) + he_1(t_n) + h^2(e_2(t_n) + \varepsilon_n^2) + \ldots + h^{N-1}(e_{N-1}(t_n) + \varepsilon_n^{N-1}) + O(h^N) \tag{4.6.5}$$

where $e_i(t)$ *are smooth functions,* $e_1(0) = 0$, *and* ε_n^j *are perturbations independent of* h *satisfying* $\varepsilon_n^{2+k} = 0$ *for* $n \geq 1+3k$, $(k = 0, 1, 2, \ldots)$. *If* B *has constant range, then the expansion exists for* $nh \leq H$, *with* H *sufficiently small, and* $\varepsilon_n^{2+k} = 0$ *for* $n \geq 1+k$.

Theorem 4.6.3 *For the index one system* (4.6.4), *the numerical solution obtained by Scheme 2 has an asymptotic expansion of the form* (4.6.5) *for* $nh \leq H$ *with* H *sufficiently small such that* $\varepsilon_n^{2+k} = 0$ *for* $n \geq 1+k$, $(k = 0, 1, 2, \ldots)$. *If* B *has constant range, then* $\varepsilon_n^2 = 0$ *for* $n \geq 1$, *and* $\varepsilon_n^{2+k} = 0$ *for* $n \geq k$ *and* $k \geq 1$.

These perturbation terms effectively impose an upper bound on the order attainable via extrapolation with either of these linearly implicit Euler methods. This upper bound depends not only on the stepsize sequence $\{n_j\}$ employed, but also on the dependence of the range of $B(y)$ in the DAE on the solution y. It is interesting to note that the behavior of a linearly implicit method, in contrast to a fully-implicit method (such as IRK or BDF), may vary significantly when the DAE is premultiplied by a nonsingular matrix dependent on y. Numerical experiments demonstrating these results are given in [171].

For the implicit Euler method applied to the index one DAE (4.6.4), there exists an expansion of the form (4.6.5) where $\varepsilon_n^j = 0$ for $n \geq 1$. Therefore, the error in the extrapolate T_{jj} is $O(H^j)$ for all j [171]. For this reason, the expense of iterating the implicit Euler method to convergence may be justified when extrapolation of the values is employed and a high order of accuracy is desired, for difficult nonlinear systems.

We note that if $B(y)$ is nonsingular as in an index zero DAE (i.e., an implicit ODE), there exists an unperturbed asymptotic error expansion for both linearly implicit schemes.

Extrapolation based on the linearly implicit midpoint method has been experimented with by Deuflhard [95] and later by Hairer and Lubich [135] for use in solving stiff ODE's. A limited theoretical result is obtained in [135] which proves the existence of an perturbed h^2 expansion of the discretization error for this method when applied to a simple stiff system of the form,

$$y' = f(y), \ y(0) = y_0$$
$$\epsilon z' = -z + g(y), \ z(0) = z_0.$$

For the corresponding DAE ($\epsilon = 0$), the pertubation terms vanish.

It is important to note that extrapolation methods do not in general enforce the satisfaction of algebraic constraints. Specifically, consider the semi-explicit index one DAE (4.6.2) obtained when $\epsilon = 0$. Of course, the first column of the extrapolation tableau, generated by the implicit Euler method, satisfies the algebraic equations. However, the extrapolated values determined via (4.6.1) do not satisfy the algebraic constraints 'exactly' (i.e., to the round-off level of accuracy). In this case it is simple, although a bit expensive, to apply extrapolation to the y variables and to compute the z variables via a Newton iteration on the algebraic constraints.

The investigation of extrapolation methods for index two DAE's is just beginning. We have seen experimental evidence in Section 4.4 for semi-explicit index two systems which indicates that extrapolation based on the implicit Euler method may be very useful. However, a complete theory justifying the use of this method is not yet available. The existence of an unperturbed asymptotic h^2 expansion of the global error for semi-explicit index two DAE's of the form

$$y' = f_0(y) + f_z(y)z$$
$$0 = g(y)$$

is shown by Lubich [172] for an extension of Gragg's method [127]. In particular, the differential equations are discretized via an explicit midpoint rule, while the algebraic equations are discretized in an implicit way. This method would be appropriate for *nonstiff* index two systems. Finally, the extrapolation method of Bader and Deuflhard [13] (namely, a semi-implicit midpoint rule) can be extended in a similar way to this class of index two systems. Lubich shows that the resulting method also has an h^2 expansion.

Chapter 5

Software for DAE's

5.1 Introduction

The most compelling reason for developing methods and analysis is to solve problems from applications. But first the methods must be implemented in codes which are efficient, robust, easy to use, and well documented. In this chapter we will discuss some of the software issues which are important in developing and using a code for solving DAE's.

Several classes of methods emerge from the convergence analysis of the previous chapters as potential candidates for methods on which to base a variable stepsize variable order general purpose code for index one DAE's. Within the class of multistep methods, the BDF methods suffer no order reduction for index one systems. One-step methods such as the L-stable singly implicit methods of Burrage and Butcher [38,39,45] and the extrapolation methods of Deuflhard et al.[96,97] which are based on the semi-implicit Euler method also appear to be promising for this purpose. The most widely used production code for DAE's at this time is the code DASSL of Petzold [200], which is based on BDF methods. We will concentrate our attention in this chapter on the algorithms and issues which are important in the implementation of DASSL. However, it is important to note that many of these issues arise in the implementation of any numerical ODE method for the solution of DAE's.

DASSL is designed to be used for the solution of DAE's of index zero and one. The convergence analysis in Chapter 3 establishes that the BDF methods achieve the same order of convergence for this class of DAE's as they do for ODE's. Thus there is some theory providing the foundation for the software. However, there are issues other than the order of convergence of the methods which are important in successfully implementing and using the methods. These issues are more complex for DAE's than for ODE's. For example, the initial conditions for a DAE must be chosen to be consistent, the linear system which must be solved on each time step is ill conditioned for small stepsizes, error estimates used in the selection of stepsizes are sensitive to inconsistencies in the initial conditions and sharp changes in the solution, and the solution methods are dependent on a more accurate approximation to the iteration matrix than is generally needed for ODE's.

Since it is possible to write a problem in the form $F(t,y,y') = 0$ which has index larger than one, it is important to be aware that a failure of the code could be due to a higher index formulation. In this case, the user of the code needs to know the possibilities of modifying the code to solve the higher index problem, or of rewriting the problem in a lower index form, and the advantages or disadvantages of these alternatives. In general, the diagnostics in DAE codes are not as well developed as those in nonstiff ODE codes. It is sometimes difficult for the code to distinguish, for example, between a failure due to inconsistent initial conditions and one due to a higher index problem formulation. Thus it is useful for anyone who plans to make more than a casual use of this type of code to be familiar with some of the details of how the code works, how it would likely behave in the event of different types of failure, and what the alternatives are for obtaining a solution in the event of code failure. On the other hand, it is our experience that the vast majority of DAE problems from applications are solved successfully with DASSL, often by users with little previous experience in solving DAE's and ODE's.

In this chapter we will explore software development issues for DAE's. In Section 5.2 we describe the basic algorithms and strategies used in DASSL, including the predictor and corrector formulas, solution of the nonlinear system, stepsize and order selection and error control. In Section 5.3 we focus on using DASSL, including how to set up a problem, how to obtain consistent initial conditions, and how to interpret failures. We also describe codes based on DASSL which are written or are currently under development with extended capabilities. It is important to remember that DASSL and other general purpose DAE codes of which we are aware are designed for solving index zero and index one DAE systems. In general, these codes will fail for higher index systems. But we have seen in Chapter 3 and Chapter 4 that the BDF methods and some other methods converge for some important classes of higher index systems, most notably for semi-explicit index two systems. In Section 5.4 we discuss the solution of higher index systems, either by rewriting the system in a lower index form which has the same analytic solution, or by solving it directly using a code like DASSL, where the error control and other strategies have been modified appropriately.

DASSL is designed for solving initial value problems of the implicit form $F(t,y,y') = 0$ which are index zero or one. At this time we are aware of several other general purpose codes for solving related problems which have been used extensively in applications. The code LSODI, developed by Hindmarsh and Painter [144], is similar to DASSL in that it is based on BDF methods. This code is written for *linearly implicit* DAE's of the form $A(t,y)y' = f(t,y)$. The user must supply a subroutine for evaluating the matrix A times a vector. LSODI differs from DASSL in the way that it implements the BDF formulas, in the way that it stores and interpolates the past solution values needed by the

BDF formulas, and most notably in the stepsize and order selection and error control strategies. Our experience with the two codes is that on many problems they are quite similar in accuracy and efficiency. For DAE's and ODE's with eigenvalues close to the imaginary axis in the complex plane, DASSL has a more robust order selection strategy. The SPRINT code, developed by Berzins, Dew and Furzeland [16,18,19,20], also employs BDF methods for the solution of linearly implicit DAE's. This code uses the filtered error estimate described in Section 5.4.2, and has been used in a wide variety of method of lines applications. The FACSIMILE code, developed by Curtis [84], uses BDF methods to solve semi-explicit index one systems. Another code of which we are aware is the code LIMEX of Deuflhard et al.[97]. This code is based on extrapolation of the semi-implicit Euler method. The theory for this method is limited to linearly implicit DAE's. We do not have any experience using this code, but Maas and Warnatz [177] report good success in solving problems in combustion modeling. This code attempts to diagnose failures in the first step, but again the technique used cannot distinguish between systems of index two (or higher) and inconsistent initial conditions for an index one problem. One-step methods, such as the extrapolation method, have an inherent advantage over multistep methods for problems with frequent discontinuities simply because they can restart after a discontinuity with a higher order method, whereas the usual implementations of multistep methods restart with a first order method.

5.2 Algorithms and Strategies in DASSL

5.2.1 Basic Formulas

DASSL is a code for solving index zero and one systems of differential/algebraic equations of the form

$$\begin{aligned} F(t, y, y') &= 0 \\ y(t_0) &= y_0 \\ y'(t_0) &= y'_0, \end{aligned} \quad (5.2.1)$$

where F, y, and y' are N-dimensional vectors. The basic idea for solving DAE systems using numerical ODE methods, originating with Gear [113], is to replace the derivative in (5.2.1) by a difference approximation, and then to solve the resulting system for the solution at the current time t_{n+1} using Newton's method. For example, replacing the derivative in (5.2.1) by the first order backward difference, we obtain the implicit Euler formula

$$F\left(t_{n+1}, y_{n+1}, \frac{y_{n+1} - y_n}{h_{n+1}}\right) = 0, \quad (5.2.2)$$

where $h_{n+1} = t_{n+1} - t_n$. This nonlinear system is then usually solved using some variant of Newton's method. We will discuss the solution of the system

(5.2.2) in the next subsection. The algorithms used in DASSL are an extension of this basic idea. Instead of always using the first order formula (5.2.2), DASSL approximates the derivative using the kth order backward differentiation formula (BDF), where k ranges from one to five. On every step it chooses the order k and stepsize h_{n+1}, based on the behavior of the solution.

DASSL uses a variable stepsize variable order *fixed leading coefficient* [147] implementation of BDF formulas to advance the solution from one time step to the next. The fixed leading coefficient implementation is one way of extending the fixed stepsize BDF methods to variable stepsizes. There are three main approaches to extending fixed stepsize multistep methods to variable stepsize. These formulations are called fixed coefficient, variable coefficient and fixed leading coefficient. The fixed coefficient methods have the property that they can be implemented very efficiently for smooth problems, but suffer from inefficiency, or possible instability, for problems which require frequent stepsize adjustments. The variable coefficient methods are the most stable implementation, but have the disadvantage that they tend to require more evaluations of the Jacobian matrix in intervals when the stepsize is changing, and hence are usually considered to be less efficient than a fixed coefficient implementation for most problems. The fixed leading coefficient formulation is a compromise between the fixed coefficient and variable coefficient approaches, offering somewhat less stability, along with usually fewer Jacobian evaluations, than the variable coefficient formulation. In contrast to DASSL, the code LSODI [144] is a fixed coefficient implementation of the BDF formulas. The relative advantages and disadvantages of the various formulations are explained in much greater detail in Jackson and Sacks-Davis [147]. It is possible that with a stepsize selection strategy which is different from the ones usually implemented in BDF codes, the variable coefficient implementation could be the most efficient [230]. However, this idea has not been extensively tested. It is still an open question for stiff ODE's and for DAE's which formulation is best for a general purpose code.

Now we will describe the basic formulas used in DASSL. Suppose we have approximations y_{n-i} to the true solution $y(t_{n-i})$ for $i = 0, 1, \ldots, k$, where k is the order of the BDF method that we are currently planning to use. We would like to find an approximation to the solution at time t_{n+1}. First, an initial guess for the solution and its derivative at t_{n+1} is formed by evaluating the *predictor polynomial* and the derivative of the predictor polynomial at t_{n+1}. The predictor polynomial $\omega_{n+1}^P(t)$ is the polynomial which interpolates y_{n-i} at the last $k+1$ times,

$$\omega_{n+1}^P(t_{n-i}) = y_{n-i}, \qquad i = 0, 1, \ldots, k.$$

The predicted values for y and y' at t_{n+1} are obtained by evaluating $\omega_{n+1}^P(t)$

5.2. ALGORITHMS AND STRATEGIES IN DASSL

and $\omega'^P_{n+1}(t)$ at t_{n+1}, so that

$$y^{(0)}_{n+1} = \omega^P_{n+1}(t_{n+1})$$
$$y'^{(0)}_{n+1} = \omega'^P_{n+1}(t_{n+1}).$$

The approximation y_{n+1} to the solution at t_{n+1} which is finally accepted by DASSL is the solution to the *corrector formula*. The formula used is the fixed leading coefficient form of the kth order BDF method. The solution to the corrector formula is the vector y_{n+1} such that the corrector polynomial $\omega^C_{n+1}(t)$ and its derivative satisfy the DAE at t_{n+1}, and the corrector polynomial interpolates the predictor polynomial at k equally spaced points behind t_{n+1},

$$\begin{aligned}
\omega^C_{n+1}(t_{n+1}) &= y_{n+1} \\
\omega^C_{n+1}(t_{n+1} - ih_{n+1}) &= \omega^P_{n+1}(t_{n+1} - ih_{n+1}), \; 1 \leq i \leq k, \\
F(t_{n+1}, \omega^C_{n+1}(t_{n+1}), \omega'^C_{n+1}(t_{n+1})) &= 0.
\end{aligned} \qquad (5.2.3)$$

In the next subsection we describe how the solution y_{n+1}, implicitly defined by conditions (5.2.3), is determined by solving a system of nonlinear equations by Newton's method.

The values of the predictor $y^{(0)}_{n+1}$, $y'^{(0)}_{n+1}$ and the corrector y_{n+1} at t_{n+1} are defined in terms of polynomials which interpolate the solution at previous time steps. Following the ideas of Krogh [155] and Shampine and Gordon [225], these polynomials are represented in DASSL in terms of modified divided differences of y. More precisely, the quantities which are updated from step to step are given by

$$\begin{aligned}
\psi_i(n+1) &= h_{n+1} + h_n + \cdots + h_{n+2-i} = t_{n+1} - t_{n+1-i}, \quad i \geq 1 \\
\alpha_i(n+1) &= h_{n+1}/\psi_i(n+1), \quad i \geq 1 \\
\beta_1(n+1) &= 1 \\
\beta_i(n+1) &= \frac{\psi_1(n+1)\psi_2(n+1)\cdots\psi_{i-1}(n+1)}{\psi_1(n)\psi_2(n)\cdots\psi_{i-1}(n)}, \quad i > 1 \\
\phi_1(n) &= y_n \\
\phi_i(n) &= \psi_1(n)\psi_2(n)\ldots\psi_{i-1}(n)[y_n, y_{n-1}, \ldots, y_{n-i+1}], \quad i > 1 \\
\phi^*_i(n) &= \beta_i(n+1)\phi_i(n), \quad i \geq 1 \\
\sigma_1(n+1) &= 1 \\
\sigma_i(n+1) &= \frac{h^i_{n+1}(i-1)!}{\psi_1(n+1)\psi_2(n+1)\cdots\psi_i(n+1)}, \quad i > 1 \\
\gamma_1(n+1) &= 0 \\
\gamma_i(n+1) &= \gamma_{i-1}(n+1) + \alpha_{i-1}(n+1)/h_{n+1}, \quad i > 1 \\
\alpha_s &= -\sum_{j=1}^{k} \frac{1}{j}
\end{aligned} \qquad (5.2.4)$$

$$\alpha^0(n+1) = -\sum_{j=1}^{k} \alpha_i(n+1).$$

The divided differences are defined by the recurrence

$$[y_n] = y_n$$
$$[y_n, y_{n-1}, \ldots, y_{n-k}] = \frac{[y_n, y_{n-1}, \ldots, y_{n-k+1}] - [y_{n-1}, y_{n-2}, \ldots, y_{n-k}]}{t_n - t_{n-k}}.$$

The predictor polynomial is given in terms of the divided differences by

$$\begin{aligned}\omega_{n+1}^P(t) = {} & y_n + (t-t_n)[y_n, y_{n-1}] + (t-t_n)(t-t_{n-1})[y_n, y_{n-1}, y_{n-2}] \\ & + \cdots + (t-t_n)(t-t_{n-1})\cdots(t-t_{n-k+1})[y_n, y_{n-1}, \ldots, y_{n-k}].\end{aligned} \quad (5.2.5)$$

Evaluating ω_{n+1}^P at t_{n+1}, and rewriting (5.2.5) in terms of the notation in (5.2.4), we obtain the predictor formula

$$y_{n+1}^{(0)} = \sum_{i=1}^{k+1} \phi_i^*(n).$$

Differentiating (5.2.5) and evaluating $\omega_{n+1}'^P$ at t_{n+1}, we find after some manipulation that

$$y_{n+1}'^{(0)} = \sum_{i=1}^{k+1} \gamma_i(n+1)\phi_i^*(n),$$

where $\gamma_i(n+1)$ satisfies the recurrence relation in (5.2.4).

To find the corrector formula, we note as in Jackson and Sacks-Davis [147] that (5.2.3) implies

$$\omega_{n+1}^C(t) - \omega_{n+1}^P(t) = b(t)(y_{n+1} - y_{n+1}^{(0)}), \quad (5.2.6)$$

where

$$b(t_{n+1} - ih_{n+1}) = 0, \quad i = 1, 2, \ldots, k$$
$$b(t_{n+1}) = 1.$$

Differentiating (5.2.6) and evaluating at t_{n+1} gives

$$\alpha_s(y_{n+1} - y_{n+1}^{(0)}) + h_{n+1}(y_{n+1}' - y_{n+1}'^{(0)}) = 0, \quad (5.2.7)$$

where the fixed leading coefficient α_s is defined in (5.2.4). Solving (5.2.7) for y_{n+1}', we find that the corrector iteration must solve

$$F\left(t_{n+1}, y_{n+1}, y_{n+1}'^{(0)} - \frac{\alpha_s}{h_{n+1}}(y_{n+1} - y_{n+1}^{(0)})\right) = 0 \quad (5.2.8)$$

for y_{n+1}.

5.2. ALGORITHMS AND STRATEGIES IN DASSL

Finally, DASSL employs an interpolant to compute the solution between mesh points. This capability is needed for output purposes, when the user requires a value of the solution at points which are not necessarily at the timesteps chosen by the code. The interpolation is a necessity in the root finding version of DASSL which we will describe later. Starting from the end of a successful step, the interpolant which DASSL uses between t_n and t_{n+1} is given by

$$\begin{aligned}\omega_{n+1}^I(t) &= y_{n+1} + (t - t_{n+1})[y_{n+1}, y_n] \\ &+ (t - t_{n+1})(t - t_n)[y_{n+1}, y_n, y_{n-1}] + \cdots \\ &+ (t - t_{n+1})(t - t_n)\cdots(t - t_{n-k+2})[y_{n+1}, y_n, \ldots, y_{n-k+1}]\end{aligned} \quad (5.2.9)$$

where k is the order of the method which was used to advance the solution from t_n to t_{n+1}. Note that the interpolant is continuous, but that it has a discontinuous derivative at t_n and t_{n+1}. It is possible to define a C^1 interpolant [15,238], which leads to a more robust code for root finding. However, at this time the C^1 interpolant is not implemented in DASSL. It will likely appear in later versions.

It is important to note that algebraic relations (constraints) for DAE's are *not* automatically satisfied at interpolated points. If this is important for the application, for example if DASSL is to be restarted at the interpolated point, then in the present version the user may need to recalculate some of the solution components at the interpolated point so that they are consistent with the constraints.

5.2.2 Nonlinear System Solution

The corrector equation (5.2.8) must be solved for y_{n+1} at each time step. Here, we describe how this is done in DASSL.

To simplify notation, rewrite the corrector equation (5.2.8) as

$$F(t, y, \alpha y + \beta) = 0 \quad (5.2.10)$$

where $\alpha = -\alpha_s/h_{n+1}$ and $\beta = y_{n+1}^{\prime(0)} - \alpha y_{n+1}^{(0)}$. In (5.2.10), all variables are evaluated at t_{n+1}, α is a constant which changes whenever the stepsize or order changes, and β is a vector which remains constant while we are solving the corrector equation.

The corrector equation is solved using a modified Newton iteration, given by

$$y^{(m+1)} = y^{(m)} - cG^{-1}F(t, y^{(m)}, \alpha y^{(m)} + \beta), \quad (5.2.11)$$

where $y^{(0)}$ is given by (5.2.5), c is a scalar constant which will be defined shortly, and G is the iteration matrix,

$$G = \alpha \frac{\partial F}{\partial y'} + \frac{\partial F}{\partial y}.$$

The partial derivatives are evaluated at the predicted y and y'.

To solve (5.2.11), the matrix G is factored into a product of an upper and lower triangular matrix, $G = LU$, and (5.2.11) is then solved by

$$\begin{aligned} Ls^{(m)} &= r^{(m)} \\ U\delta^{(m)} &= s^{(m)}, \end{aligned} \qquad (5.2.12)$$

where $\delta^{(m)} = y^{(m+1)} - y^{(m)}$, and $r^{(m)} = -cF(t, y^{(m)}, \alpha y^{(m)} + \beta)$. In DASSL, the matrix G may be dense or have a banded structure. The factorization of G and the solution of the systems in (5.2.12) are performed by routines in the LINPACK [100] software package.

For many applications, and especially when the system to be solved is large, the costs of computing and factoring G dominate the cost of the integration. Often the matrices $\partial F/\partial y'$ and $\partial F/\partial y$ change very little over the span of several time steps. However, the constant α changes whenever the stepsize or order of the method being used changes. If the derivative matrices and the constant α have not changed very much since the last iteration matrix was computed, then it is desirable to use the old iteration matrix in (5.2.11), instead of reevaluating the matrix on every step. Thus, if $\hat{G}$ is the iteration matrix that we have saved from some previous time step, $\hat{G} = \hat{\alpha}\partial F/\partial y' + \partial F/\partial y$, where $\hat{\alpha}$ depends on the stepsize and order at some past time step when $\hat{G}$ was last computed, then the iteration we actually use is

$$y^{(m+1)} = y^{(m)} - c\hat{G}^{-1}F(t, y^{(m)}, \alpha y^{(m)} + \beta). \qquad (5.2.13)$$

As long as $\hat{G}$ is close enough to G, (5.2.13) will converge at an adequate rate, and we will be able to solve the corrector equation.

The constant c is chosen to speed up the rate of convergence of the iteration when $\alpha \neq \hat{\alpha}$. The idea of using this type of acceleration was introduced in [40,156]. Choosing $c \equiv 1$ corresponds to the usual modified Newton iteration for solving (5.2.10) for y, and choosing $c \equiv \alpha/\hat{\alpha}$ corresponds to the modified Newton iteration for solving $F(t, \hat{\alpha}^{-1}(y' - \beta), y') = 0$, (where $y' = \hat{\alpha}y + \beta$) for y'. To find the optimal value of c, note that the iteration (5.2.13) converges, for a sufficiently good initial guess, when the spectral radius of

$$B = (I - c\hat{G}^{-1}G)$$

is less than one. When (5.2.13) is used to solve the linear ODE, $y' - Ay = 0$, we find that if λ is an eigenvalue of A, then $\bar{\lambda} = 1 - c(\alpha - \lambda)/(\hat{\alpha} - \lambda)$ is an eigenvalue of B. For a general DAE system where the eigenvalues of G are unknown, we take c to minimize the maximum of this expression over all λ in the left half of the complex plane. This minimization gives

$$c = \frac{2\hat{\alpha}}{\alpha + \hat{\alpha}}.$$

5.2. ALGORITHMS AND STRATEGIES IN DASSL

When c is chosen in this way, the corresponding $\bar{\lambda}$ is

$$\bar{\lambda} = \frac{\hat{\alpha} - \alpha}{\alpha + \hat{\alpha}}. \qquad (5.2.14)$$

The expression (5.2.14) is useful in deciding when to compute a new iteration matrix. Assuming that $\bar{\lambda}$ is an approximation to the rate of convergence of (5.2.13) when $\alpha \neq \hat{\alpha}$, and that we are willing to tolerate a rate of convergence of μ (in DASSL, $\mu = 0.25$), then a new iteration matrix is computed before attempting the iteration whenever the stepsize and/or order changes sufficiently so that $|\bar{\lambda}| > \mu$. Of course, a new iteration matrix is also computed whenever the corrector fails to converge after four iterations.

A decision which is important to the reliability and efficiency of a code like DASSL is when to terminate the corrector iteration (5.2.11). It is well-known that

$$\|y^* - y^{(m+1)}\| \leq \frac{\rho}{1-\rho}\|y^{(m+1)} - y^{(m)}\|$$

where y^* is the solution to the corrector equation and ρ is an estimate of the rate of convergence of the iteration. Following Shampine [221], the iteration is continued until

$$\frac{\rho}{1-\rho}\|y^{(m+1)} - y^{(m)}\| < 0.33 \qquad (5.2.15)$$

so that the iteration error $\|y^* - y^{(m+1)}\|$ will be sufficiently small. While condition (5.2.15) appears at first to be independent of any user requested error tolerances, this information is buried in the norm. In particular, the default norm used in DASSL is a weighted root mean square norm, where the weights depend on the relative and absolute error tolerances and on the values of y at the beginning of the step. The constant 0.33 was chosen so that the errors due to terminating the corrector iteration would not adversely affect the integration error estimates. More precisely, if the weights are all equal and proportional to a desired integration error tolerance ϵ, then condition (5.2.15) may be interpreted as restricting the corrector iteration error to be less than $.33\epsilon$. A maximum of four iterations is permitted. Whenever $\rho > 0.9$, the corrector iteration is considered to have failed, and the iteration matrix is reevaluated if it isn't current.

It is possible for the user to define a personalized norm by substituting a routine for computing the norm. This option is important in some applications, for example in the method of lines solution of partial differential equations.

The rate of convergence is estimated whenever two or more corrector iterations have been taken by

$$\rho = \left(\frac{\|y^{(m+1)} - y^{(m)}\|}{\|y^{(1)} - y^{(0)}\|}\right)^{1/m}.$$

If the difference between the predictor and the first correction is very small (relative to roundoff error in y), the iteration is terminated, unless α has changed or a new iteration matrix has just been formed. In these last two cases, a second corrector iteration is forced and a new estimate of ρ will be computed. If the iteration is terminated after only one corrector iteration, ρ is not recomputed. This last exception is needed because the rate may be polluted by roundoff error, and hence be misleading. On other steps, the rate of convergence is taken to be the last rate computed.

If the corrector iteration fails to converge, the iteration matrix is reevaluated and the iteration is attempted again. With a new iteration matrix, if the corrector again fails to converge, the stepsize is reduced by a factor of one quarter. After ten consecutive failures, or if the stepsize becomes so small that $t+h$ is equal to t in computer arithmetic, DASSL returns to the main program with an error flag.

The matrix G is either computed by finite differences, or supplied directly by the user. In finite differencing, if G is banded, the columns of G having nonzero elements in different rows are approximated simultaneously, using an idea of Curtis et al.[85]. The jth column of G is approximated by

$$\frac{F(t, y + \delta_j e_j, \alpha(y + \delta_j e_j) + \beta) - F(t, y, \alpha y + \beta)}{\delta_j}, \qquad (5.2.16)$$

where e_j is the jth unit vector, and δ_j is the increment. This formulation of the differencing uses fewer function evaluations than differencing $\partial F/\partial y$ and $\partial F/\partial y'$ separately, and then adding them together to form $G = \alpha \partial F/\partial y' + \partial F/\partial y$.

The differencing procedure can be sensitive to roundoff error. The main source of difficulty in computing the iteration matrix G by finite differencing is the choice of the increment δ_j. The choice

$$\delta_j = \text{sign}(h y_j') \max(|y_j|, |h y_j'|, \text{WT}_j) \sqrt{u}, \qquad (5.2.17)$$

where $\text{WT}_j = \text{RTOL}|y_j| + \text{ATOL}$ is the weight vector used in the norm, is used in DASSL. RTOL and ATOL are the relative and absolute error tolerances specified by the user, and u is the unit roundoff error of the computer. Normally, when $|y_j|$ is not very close to zero, about half of the digits of y will be perturbed by the increment determined by (5.2.17). If $|y_j|$ is near zero, it is quite possible that a nearby value of y_j is not so small, and since $y_j(t+h) \approx y_j + h y_j'$, $|h y_j'|$ was included in the maximum in (5.2.17) to prevent a near zero perturbation from being selected. In the event that $|y_j|$ and $|h y_j'|$ are both near zero, WT_j is included in the maximum because the user has told us implicitly by setting the error tolerances that it is the smallest number which is relevant with respect to y_j. It is important to note that the sign of the increment will be negative if the solution is decreasing. This choice can occasionally be a source of difficulty for problems where F is undefined for $y < 0$, or not differentiable at $y = 0$.

5.2.3 Stepsize and Order Selection Strategies

An important feature of DASSL is that it changes the stepsize and the order of the method to solve problems more efficiently and reliably. In this subsection we describe the criterion used to decide when to accept a step, and the strategies for selecting the stepsize and order for the next step.

One of the most important questions concerning reliability of a DAE code is how it decides to accept or reject a step. There are two sources of errors that we are concerned with. The first is the local truncation error of the method. It is the amount by which the solution to the DAE fails to satisfy the BDF formula. The fixed leading coefficient form of the BDF approximates hy' with the formula

$$h_{n+1}y'_{n+1} = \sum_{i=2}^{k+1} \alpha_{i-1}(n+1)\phi_i(n+1) - (\alpha_s - \alpha^0(n+1))\phi_{k+2}(n+1),$$

where the notation is defined in equation (5.2.4). The true solution $y(t)$ satisfies

$$h_{n+1}y'(t_{n+1}) = \sum_{i=2}^{k+1} \alpha_{i-1}(n+1)\tilde{\phi}_i(n+1) \\ -(\alpha_s - \alpha^0(n+1))\tilde{\phi}_{k+2}(n+1) + \tau_{n+1},$$

where τ_{n+1} is the local truncation error, given by

$$\tau_{n+1} = \left(\alpha_{k+1}(n+1) + \alpha_s - \alpha^0(n+1)\right)\tilde{\phi}_{k+2}(n+1) \\ + \sum_{i=k+3}^{\infty} \alpha_{i-1}(n+1)\tilde{\phi}_i(n+1), \tag{5.2.18}$$

and $\tilde{\phi}_i(n+1)$ are the scaled divided differences (i.e, the ϕ_i) defined in (5.2.4), evaluated at the true solution $y(t)$. DASSL estimates the principal term of the local truncation error, which is the first term in (5.2.18).

The divided difference $\tilde{\phi}_{k+2}(n+1)$ in (5.2.18) is approximated by $\phi_{k+2}(n+1)$, and $\phi_{k+2}(n+1)$ is the same as the difference between the predictor and the corrector. To see this, let $\omega^*_{n+1}(t)$ be the unique polynomial that interpolates y_{n+1-i}, $i = 0, 1, \ldots, k+1$. Then recalling that the predictor interpolates y_{n-i}, $i = 0, 1, \ldots, k$, it follows directly from the equations for the interpolating polynomials that $\omega^*_{n+1}(t) - \omega^P_{n+1}(t) = \phi_{k+2}(n+1)$ at $t = t_{n+1}$. The desired result follows immediately since $\omega^*_{n+1}(t_{n+1}) = y_{n+1}$ and $\omega^P_{n+1}(t_{n+1}) = y^{(0)}_{n+1}$ by definition.

This estimate of $\tilde{\phi}_{k+2}(n+1)$ is asymptotically correct in the case of constant stepsize or under slightly more general conditions [115,147]. The stepsize and order selection strategies in DASSL tend to favor sequences of constant stepsize and order, and the estimate has worked quite well in practice. Thus, the principal term of the local truncation error is estimated by

$$\left(\alpha_{k+1}(n+1) + \alpha_s - \alpha^0(n+1)\right) \|y_{n+1} - y^{(0)}_{n+1}\|. \tag{5.2.19}$$

The other source of error that is important is the error in interpolating to find the solution between mesh points for output purposes. In DASSL the solution at a point t^* between mesh points, $t_n \leq t^* \leq t_{n+1}$, is found by evaluating the polynomial of degree k (where k is the order of the method which was used to find y_{n+1}) which interpolates y at the last $k+1$ points $y_{n+1}, y_n, \ldots, y_{n-k+1}$, at t^*. The error in this interpolated value is composed of two parts. The first part is the *interpolation error*, due to interpolating the true solution $y(t)$ at $t_{n+1}, t_n, \ldots, t_{n-k+1}$ by a polynomial. The principal term of the interpolation error is bounded by

$$\alpha_{k+1}(n+1)\|\phi_{k+2}(n+1)\|. \tag{5.2.20}$$

The second part of the error is due to the fact that we are interpolating the computed solution instead of the true solution. This error is proportional to the errors in the solution at the mesh points. We are already trying to control these errors in (5.2.19), so we do not approximate them here.

Taking (5.2.19) and (5.2.20) together, at every step we require that

$$\text{ERR} = M\|y_{n+1} - y_{n+1}^{(0)}\| \leq 1.0, \tag{5.2.21}$$

where

$$M = \max\left(\alpha_{k+1}(n+1), |\alpha_{k+1}(n+1) + \alpha_s - \alpha^0(n+1)|\right).$$

If this condition is not satisfied, then the step is rejected. Note that condition (5.2.21) requires that the estimated integration error be less than the requested integration error tolerances, as reflected in the weighted norm of the predictor-corrector difference.

Whether or not a step is rejected, the code must decide which order method is to be used on the next step. Since the predictor (5.2.5) and the corrector (5.2.7) are of the same order, the leading order terms in their corresponding Taylor series expansions are identical aside from the error constants. DASSL estimates the leading order term in the remainder of the Taylor series expansion, independent of the error constants, at orders $k-2$, $k-1$, and k. If the last $k+1$ steps have been taken at constant stepsize and order, then the leading error term for order $k+1$ is also estimated. These estimates are

$$\begin{aligned}
\text{TERKM2} &= \|(k-1)\sigma_{k-1}(n+1)\phi_k(n+1)\| \approx \|h^{k-1}y^{(k-1)}\| \\
\text{TERKM1} &= \|k\sigma_k(n+1)\phi_{k+1}(n+1)\| \approx \|h^k y^{(k)}\| \\
\text{TERK} &= \|(k+1)\sigma_{k+1}(n+1)\phi_{k+2}(n+1)\| \approx \|h^{k+1}y^{(k+1)}\| \\
\text{TERKP1} &= \|(k+2)\sigma_{k+2}(n+1)\phi_{k+3}(n+1)\| \approx \|h^{k+2}y^{(k+2)}\|.
\end{aligned} \tag{5.2.22}$$

Since the higher order estimate TERKP1 is formed only after $k+1$ steps at constant stepsize, $\sigma_{k+2}(n+1) = 1/(k+2)$. Note that these estimates are scaled to be independent of the error constants for each order. The strategies for raising or lowering the order are nearly identical with those described in

5.2. ALGORITHMS AND STRATEGIES IN DASSL

Shampine and Gordon [225]. Briefly, there is an initial phase where the order and stepsize are increased on every step, and after that the order is raised or lowered depending on whether TERKM2, TERKM1, TERK and TERKP1 form an increasing or decreasing sequence. The philosophy behind this type of order selection strategy is that the Taylor series expansion is behaving as expected for higher orders only if the magnitudes of successive higher order terms form a decreasing sequence. Otherwise, it is safer to use a lower order method.

It is interesting to note that the order selection strategy in DASSL differs from that used in the LSODE family of codes [144] in several important aspects. In LSODE and its variants, error estimates are formed for the order $k-1$, k and $k+1$ methods, as if the last $k+1$ steps were taken at constant stepsize. The order is chosen which gives the largest stepsize based on these estimates. Thus, the order is chosen based solely on efficiency considerations. In DASSL, the terms $||h^{k-1}y^{(k-1)}||$, $||h^k y^k||$, etc. in the Taylor series expansion, rather than the error estimates for those orders, are compared. Several successive terms are examined, and the order is chosen based on maintaining a decreasing sequence. While the terms that DASSL compares are related to the error estimates, they are not the same.

The advantage of the order selection strategy which is implemented in DASSL is that it effectively solves a problem that variable order BDF codes can have because the higher order BDF methods are unstable for some ODE's. Consider the model problem $y' = Ay$, where A has eigenvalues with large imaginary parts but small negative real parts. The higher order BDF methods are unstable for this type of problem unless the stepsize is chosen to be quite small. If these small stepsizes are not required for the resolution of the solution, then the BDF code can take larger stepsizes with the more stable, lower order formulas than with the higher order formulas. The presence of instability is evidenced by rapid oscillations, or more simply by $||h^p y_n^{(p)}|| > ||h^q y_n^{(q)}||$ for $p > q$ [231]. When DASSL is solving a problem of this type and the higher order formulas begin to have a problem with instability, the estimates TERKM2, TERKM1, TERK, and TERKP1 fail to form a decreasing sequence. Then, the order is lowered until these terms (recomputed for the new order) are again decreasing. Thus the code is forced to use a stable, lower order method, thereby allowing larger stepsizes. The same sort of phenomena occurs with the error estimates in LSODE, but in DASSL testing for monotonicity and comparing the terms in (5.2.22) rather than the error estimates favors the lower order methods and force a lowering of the order sooner. The strategy used in the LSODE codes sometimes fails to lower the order, and in those cases the code can be quite inefficient because it continues to use the higher order BDF and needs to take very small stepsizes to maintain stability. It is important to note that for all modern ODE solvers, the stepsize selection strategy reduces the stepsize as a response to instability. Thus the codes do not give a faulty solution because of the instability for large stepsizes, but instead reduce the stepsize and become very inefficient. Because of these differences in the order selection strategies between DASSL and LSODE, DASSL is more robust than the LSODE codes for this type of problem. Order selection strategies to control instability in higher order BDF methods are discussed in detail in [231].

After the step has been accepted or rejected and the order of the method to be used on the next step has been determined, DASSL decides what stepsize to use on the next step. The most effective strategy, based on our experiences and experimentation, is the one used by Shampine and Gordon [225]. The error at the new order k is estimated as if the last $k+1$ steps were taken at the new stepsize, and the new stepsize is chosen so that the error estimate satisfies (5.2.21). More precisely, DASSL chooses the new stepsize rh_{n+1} conservatively so that the error is roughly one half of the desired integration error tolerance. Thus r is given by

$$r = (2.0 \text{EST})^{-1/(k+1)}, \qquad (5.2.23)$$

where EST is the error estimate for the order k method which was selected for the next step and is given by the term estimated in (5.2.22) for the new order k, but scaled now by the error constant $1/(k+1)$. After a successful step, the stepsize is increased only if it can be doubled, and then it is doubled. If a decrease is called for after a successful step, the stepsize is decreased by at least a factor $r = 0.9$, and at most $r = .5$. After an unsuccessful step, if this is the first failure since the last successful step, then r is determined by (5.2.23) and multiplied by 0.9. The stepsize is reduced by at least $r = 0.9$ and at most $r = 0.25$. After the second failure, the stepsize is reduced by a factor of one quarter, based on the philosophy that the error estimates can no longer be trusted.

DASSL keeps a count of the number of integration error test failures since the last successful step. After a large stepsize decrease, the fixed leading coefficient BDF formulas are most accurate at order one, in contrast to the variable coefficient formulation where the accuracy does not deteriorate so much for higher order formulas when the stepsize is decreased drastically. It is likely that the first order formula is best anyway in this situation because the past values of y will not yield much useful information if the error test has already failed several times. After three consecutive error test failures, the order is reduced to one, and the stepsize is reduced by a factor of one quarter on every failure thereafter. When the stepsize is so small that $t + h \approx t$, an error return is made. In DASSL this minimum stepsize is computed as $h_{\min} = 4u \max(|t_n|, |TOUT|)$ where u is the unit roundoff error, t_n is the current mesh point, and TOUT is the user requested output point.

As a final issue in this subsection, we discuss how DASSL selects the initial stepsize. Even for ODE codes, this issue is somewhat tricky. Numerous strategies have been proposed. See for example [116,225,237]. DASSL uses

$$h_0 = \text{sign}(\text{TOUT} - \text{T}) \min\left(10^{-3}|\text{TOUT} - \text{T}|, \frac{1}{2}\|y'\|^{-1}\right).$$

5.3. OBTAINING NUMERICAL SOLUTIONS

This strategy is designed to yield a successful step for a zeroth order method and is quite conservative. If $||y'||$ is very small, then it is hoped that TOUT−T gives some information on the scale of the solution interval. It is not terribly uncommon in applications for $||y'||$ to be zero or very small at the initial time, as frequently the derivative values are unknown for the algebraic variables. In these cases, the value of TOUT influences the initial stepsize. Hence, different values of TOUT can change the performance of the code. This aspect of DASSL's initial stepsize algorithm is similar to the strategy proposed by Söderlind in his code DASP3 [233] for the numerical solution of coupled stiff ODE's and DAE's. There is an option in DASSL for the user to select the initial stepsize.

5.3 Obtaining Numerical Solutions

5.3.1 Getting Started with DASSL

DASSL is designed to be as easy to use as possible, while providing enough flexibility and control for solving a wide variety of problems. It is extensively documented in the source code. The user interface for DASSL is based on the user interface for ODE solvers proposed by Shampine and Watts in [226], with a few changes which are necessary to accommodate the more general DAE systems. In this subsection we outline what a user must do to solve a problem with DASSL.

We emphasize that DASSL is designed for solving index zero and index one problems of the form

$$
\begin{aligned}
F(t, y, y') &= 0 \\
y(t_0) &= y_0 \\
y'(t_0) &= y'_0,
\end{aligned}
\quad (5.3.1)
$$

where F, y and y' are N-dimensional vectors. DASSL makes use of a subroutine RES which is written by the user to define the function F in (5.3.1). RES takes as input the time T and the vectors Y and YPRIME, and produces as output the vector DELTA, where DELTA = F(T,Y,YPRIME) is the amount by which the function F fails to be zero for the input values of T, Y and YPRIME. The subroutine has the form

SUBROUTINE RES(T,Y,YPRIME,DELTA,IRES,RPAR,IPAR)

The parameter IRES is an integer flag which DASSL always sets to zero prior to calling RES. It is used to flag situations where an illegal value of Y or a stop condition has been encountered. For example, in some applications, if a component of Y becomes even slightly negative, then the function F cannot be evaluated. In this case, the user would check for a negative component of Y upon entering RES. If one is found, then the user would set IRES = −1 and return without evaluating the function. DASSL then cuts the stepsize and attempts the step again. RPAR and IPAR are real and integer vectors, respectively, and are at the user's disposal to use for communication purposes. They are never altered by DASSL or any of its subroutines.

To get started, DASSL needs a consistent set of initial values T, Y and YPRIME. A necessary but not always sufficient condition for consistent initialization is that F(T,Y,YPRIME) = 0 at the initial time. At the time of this writing, we are not aware of any general codes for computing consistent initial values of Y and YPRIME, given enough information about Y and YPRIME to specify a unique solution to the analytical problem. An algorithm for accomplishing this task in general has recently been developed and will be discussed in Section 5.3.4. For problems in which the initial values of Y are given, there is an option in DASSL to compute the initial value of YPRIME, given a starting guess for YPRIME. In this case, DASSL takes a small implicit Euler step for its first step, and uses a damped Newton iteration to solve the nonlinear system. The error estimate on this step is different from the estimate which DASSL usually uses because the initial derivatives are not available for use in an error estimate. It is sometimes possible to start DASSL without consistent values of YPRIME, but it should be noted that the error estimates on the first step are not correct in this case unless the user has specified the option for the code to compute the initial values of YPRIME.

The call to DASSL is

CALL DASSL(RES,NEQ,T,Y,YPRIME,TOUT,INFO,RTOL,ATOL,
IDID,RWORK,LRW,IWORK,LIW,RPAR,IPAR,JAC)

The parameters are described in detail in the documentation to DASSL, so we will only discuss a few features here. Many of these features are activated by setting an element of the option vector INFO to one. The user subroutines RES and JAC must be declared external in the user's main program. The variable NEQ equals N, the number of equations in the system (5.3.1). In case DASSL fails, the scalar variable IDID should be examined to see what specific error caused the difficulty. The documentation of DASSL gives a detailed explanation of all the error messages.

Frequently a user desires the numerical solution at a set of output times, so the usual way to call DASSL is in a loop which increments the output time TOUT until the end of the interval has been reached. The value of the solution which DASSL returns is the value of the interpolant (5.2.9) at those times. To obtain more detailed information about the solution at the internal timesteps that DASSL selects, there is an option to have DASSL return after each internal time step.

As we have described in Section 5.2.2, the algorithms in DASSL require an estimate of the iteration matrix, $G = \alpha \partial F/\partial y' + \partial F/\partial y$. There is an option to specify this matrix directly which requires the user to provide a subroutine JAC. The default is for DASSL to approximate this matrix via finite

5.3. OBTAINING NUMERICAL SOLUTIONS

differences. The advantage of the finite difference option is that it is simpler. We strongly recommend, especially for novice users and new problems, to let DASSL compute the matrix. The JAC subroutine is usually not simple to write and debug. The consequence of a very poor Jacobian matrix approximation is a serious deterioration of efficiency or possibly even complete failure. For cases in which DASSL is inefficient because of a poor finite difference Jacobian approximation, providing a subroutine JAC will improve performance. In Section 5.3.3, we give some suggestions on how to recognize this situation. Which option is most efficient is quite problem dependent.

For many problems in applications, the iteration matrix is banded. If the code is informed of this matrix structure by an appropriate input, then the storage needed will be greatly reduced, the numerical differencing will be performed much more cheaply, and a number of important algorithms will execute much faster. In the case that the user computes a banded matrix via subroutine JAC, the elements must be stored in a special format which is explained in the documentation.

For a very few problems in applications whose analytic solutions are always positive, we have found that it is crucial to avoid small negative solutions which can arise due to truncation and roundoff errors. There is an option in DASSL to enforce nonnegativity of the solution. However, we strongly recommend against using this option except when the code fails or has excessive difficulty without it.

Finally, a comment on the relative and absolute error tolerances RTOL and ATOL. These variables can be specified either as vectors or as scalars. Most of the important decisions in the code make use of these tolerances to compute weights for the norm, and the finite difference Jacobian approximation makes use of them directly when there is no other information available about the scale of the solution. *We cannot emphasize strongly enough the importance of carefully selecting these tolerances to accurately reflect the scale of the problem.* In particular, for problems whose solution components are scaled very differently from each other, it is advisable to provide the code with vector valued tolerances. For users who are not sure how to set the tolerances RTOL and ATOL, we recommend starting with the following rule of thumb. Let m be the number of significant digits required for solution component y_i. Set $RTOL_i = 10^{-(m+1)}$. Set $ATOL_i$ to the value at which $|y_i|$ is essentially insignificant.

The norm which DASSL uses is a weighted root mean square norm, given by

$$||\mathbf{v}|| = \sqrt{(1/\text{NEQ}) \sum_{i=1}^{\text{NEQ}} (v_i/\text{WT}_i)^2}$$

where

$$\text{WT}_i = \text{RTOL}_i |Y_i| + \text{ATOL}_i.$$

These are the best choices for the majority of problems. However, for some problems, particularly method of lines solutions to partial differential equations, it is better to make use of additional information on the structure of Y to define the weights and/or the norm. It is possible for the user to replace these subroutines, but we generally recommend against this. Details are given in the DASSL documentation.

The FORTRAN source code for DASSL is available by way of the National Energy Software Center (NESC). The address is Argonne National Laboratory, 9700 South Cass Avenue, Argonne, Illinois 60439, U.S.A. Single and double precision versions are available. The accession number for DASSL at NESC is 9918. Prospective users from Western Europe or Japan can obtain DASSL from the NEA Data Bank. The address is B.P. No. 9 (Bat. 45) F-91190, GIF-sur-YVETTE, France.

5.3.2 Troubleshooting DASSL

In our experience, DASSL solves most DAE systems from applications without difficulty. However, we have occasionally encountered systems for which DASSL has trouble. In this subsection we will describe some of the possible sources of trouble and give some advice on how to recognize the problem, along with possible remedies.

In diagnosing problems, it is helpful to be aware of the existence of information about the internal operations of DASSL. Information which is accessible through the work arrays RWORK and IWORK includes: the stepsize to be attempted on the next step, the stepsize used on the last successful step, the order of the method to be attempted on the next step, the order of the method used on the last step, and running totals of the number of steps taken, the number of calls to RES, the number of evaluations of the iteration matrix needed, the number of integration error test failures, and the number of convergence test failures. We have found the information on integration error test failures and convergence test failures to be particularly useful. The locations of this information are given in the DASSL documentation.

Before proceeding, we wish to emphasize that the DASSL code has successfully solved a wide variety of scientific problems and has had a large number of users. A user who suspects a bug should first very carefully examine his own code and reread the DASSL documentation. *In our experience, bugs in user code and incorrect use of DASSL are by far the most likely causes of difficulties.*

Inaccurate Numerical Jacobian

The finite difference Jacobian calculation is, in our opinion, the weakest part of the DASSL code. It is a difficult problem to devise an algorithm for calculating the increments for the differencing which is both cheap and robust. We have seen cases of extremely poorly scaled systems from applications, where some elements of the finite difference Jacobian were computed to be zero when they

5.3. OBTAINING NUMERICAL SOLUTIONS

should have been quite large. To see how this can happen, suppose there is a term in the system which looks like $\lambda(y_1 - y_2)$, where $y_1 \gg y_2$. In computing the partial derivative of this term with respect to y_2, the difference is based on the size of y_2. Thus if y_1 is large enough compared to y_2, we will have $y_1 - (y_2 + \delta y_2) = y_1 - y_2$ in machine arithmetic and the code will think that the partial derivative is zero. It is likely that in future versions of DASSL a more robust finite difference Jacobian algorithm such as the one proposed by Salane [218] will be considered.

Usually, the symptoms of an inaccurate numerical Jacobian are a stepsize that is much smaller than appears to be warranted from accuracy considerations, and high ratios of convergence test failures both to error test failures and to steps taken so far.

There are several possible remedies to difficulties with obtaining good finite difference approximations to the Jacobian matrix. The function subroutine RES can often be rewritten or a change of variables made to avoid cancellations and poorly scaled variables. A user defined analytic JAC routine can be written, or a user defined finite difference Jacobian routine can be substituted. Sometimes a more careful look at the error tolerances helps. In particular, increasing ATOL for exceedingly small components which do not need to be computed accurately will often remedy this problem. This is because of the way that ATOL is used in the calculation of the increment δ_j for finite difference Jacobians, as we have explained in Section 5.2.2.

Finally, we note that if there is an iteration inside the user RES routine (for example, to solve a nonlinear system), then the increment for the numerical Jacobian should be based on the error of this iteration, rather than on the unit roundoff error. Because such an iteration can also cause difficulties with the stepsize selection and error estimates in DASSL, we strongly encourage DASSL users to avoid such situations whenever possible by reformulating the problem.

Inconsistent Initial Conditions/Discontinuities

If the user-supplied initial conditions for y and y' are inconsistent, then DASSL may fail on the first step. The symptom of the failure can be either successive error test failures or, for nonlinear problems, the Newton iteration may fail to converge due to a poor initial guess. For index one systems with inconsistent initial conditions, the error estimate tends to a constant as the stepsize approaches zero. To see why this is true, consider the simple algebraic equation $y = g(t)$, with $y_0 = a \neq g(t_0)$. DASSL always starts with the implicit Euler method because it requires no memory of past steps. An implicit Euler step for this system gives $y_1 = g(t_1)$. The error estimate is $(1/2)\|y_1 - y_1^{(0)}\|$, which tends to $(g(t_0) - a)$ as $t_1 \to t_0$. A more extensive discussion of the difficulties encountered by codes based on the BDF methods when applied to DAE's is given in [200].

It is important to note that initial guesses which are only slightly inconsistent can cause DASSL to fail to complete the first step. Failure occurs because the errors, and therefore the inconsistencies, are measured in the weighted norm which is determined by RTOL and ATOL. Hence it is important for the user to compute these initial conditions very accurately. If a discontinuity is introduced, perhaps inadvertently by a user input function, into an algebraic variable of the system, this same behavior can also occur in later steps.

Higher Index Systems

DASSL cannot solve higher index systems without modifications to the error estimates and other strategies in the code. We will discuss the possibilities for solving higher index systems in Section 5.4. Here we focus on recognizing these types of systems when they have inadvertently been given to DASSL. In general, the error estimates used in DASSL do not tend to zero as the current stepsize tends to zero for systems whose index is greater than one – in fact, they may even diverge! As a response to a large integration error estimate, the code drastically (and repeatedly) reduces the stepsize until the iteration matrix becomes ill-conditioned. For a system of index ν, we show in Section 5.4.2 that the iteration matrix has condition $O(h^{-\nu})$. For small enough stepsizes, this conditioning problem causes the Newton iteration not to converge, and the code eventually fails due to multiple convergence test failures. There is an error return flag in DASSL which signals multiple integration error test failures followed by multiple convergence test failures. This error return is usually a signal of a higher index problem, or possibly of an index one problem with inconsistent initial conditions or a discontinuity. However, it is also quite possible for these types of problems to fail solely because of multiple convergence test failures, particularly for nonlinear problems with index greater than two. Sometimes DASSL can start solving a smooth higher index system and fail after a stepsize or order change or a rapid change in the solution, so the failures will not necessarily occur on the first step.

Singular Iteration Matrix

While it is possible to define a solvable DAE system whose iteration matrix is singular for all possible values of the stepsize, we have never seen such a problem occur in an application and it would necessarily have to have an index which was greater than one. DASSL gives an error return flag for a singular iteration matrix only after the code has tried to compute the iteration matrix for several quite disparate stepsizes and has always produced a matrix which is singular. It is possible that a singular matrix can be created by a very inaccurate finite differencing of the matrix within DASSL or by a bug in a user supplied JAC routine. Singularity can also occur because there are redundant equations in the system to be solved (that is, redundant equations in RES). In this case the user needs to reformulate the problem to include

5.3. OBTAINING NUMERICAL SOLUTIONS 135

some extra information and remove the redundant information. Otherwise there is no possibility for obtaining a unique solution. In our experience, this bug is a fairly common occurrence in codes using DASSL in the method of lines solution of partial differential equations when the boundary conditions have been improperly specified.

Poorly Scaled System/Inappropriate Error Tolerances

The error tolerances RTOL and ATOL are the means by which the user communicates to DASSL the accuracy requirements of the solution and the scale of the problem. It is possible for DASSL to miss important changes in the solution because these error tolerances are too loose. On the other hand, if the tolerances are too tight, then DASSL will work very hard to get accurate solutions for variables which may be very small and insignificant. In particular, we do not recommend using pure relative error control (ATOL = 0) for solution components which may be small or become small during the integration. The code tends to function most efficiently if it does not need to take unnecessarily small stepsizes because the iteration matrix becomes more singular as the stepsize tends to zero. On the other hand, the error estimates in DASSL are valid only for sufficiently small stepsizes. As with many ODE codes, DASSL may not perform well with exceedingly loose tolerances. We recommend asking for at least two digits of accuracy except possibly for exceedingly small and unimportant solution components.

5.3.3 Extensions to DASSL

Although DASSL is a powerful code which can handle a wide variety of problems, some problems require capabilities which are not implemented in the standard version of DASSL or which can be performed more efficiently with a code based on DASSL but oriented towards a more specific task. In this subsection we will discuss DASSLRT, a root finding version of DASSL, DASSAC, a version of DASSL for sensitivity analysis problems, and several other extensions to DASSL which are at this time under development.

The standard version of DASSL solves a DAE from time T to time TOUT, where TOUT is specified by the user. In some problems it is more natural to stop the code at the root of some function $g(t,y)$. For example, in computing the trajectory of an object, it does not make sense to continue the solution past the time when the object hits the ground. In other problems, it is necessary to stop the code at the root of g because the function F changes at the roots of g. For example, in computing the flow of a fluid, the flow is 'choked' if the pressure ratio exceeds a critical value, and unchoked otherwise. Choked flows obey a different set of DAE's than unchoked flows, so it is necessary to check for the root of the function defined by the pressure ratio minus the critical value, and define the function F in RES according to whether the flow is currently choked or not.

In our experience, some users have attempted to implement these types of root finding problems by first inserting an IF statement in the RES routine and switching the function based on the sign of g. We do not, in general, recommend this approach. Since linear multistep methods utilize function and derivative information over several integration steps, it does not make sense to switch the function (effectively introducing a discontinuity in the derivatives) without restarting the integration. The sudden change in the derivatives of F can wreak havoc with the numerical Jacobian approximation, Newton iteration, error estimates, and stepsize and order selection algorithms in DASSL. It can be grossly inefficient and even lead to failure for some problems. Instead, we recommend setting a flag in the main program which determines which case of the function is to be evaluated. The flag is communicated to RES via IPAR. Then, DASSLRT should be called. When it returns at a root, the flag is changed in the main program. It may also be necessary at this time to enforce consistency of the constraint equations. Then DASSLRT is restarted. For complex problems which involve several cases of F, it may be necessary to use several flags.

It is helpful in formulating problems to understand roughly how DASSLRT works. First, whenever DASSL completes a successful step to t_{n+1}, it evaluates the root function $g(t_{n+1}, y_{n+1})$. If the sign of $g(t_{n+1}, y_{n+1})$ is different from the sign of $g(t_n, y_n)$, then a root must have occurred in the interval $[t_n, t_{n+1}]$. DASSL then calls a root solver to find the root of $g(t, \omega_{n+1}^I(t)) = 0$ in $[t_n, t_{n+1}]$. The root solver in DASSLRT is based on ideas developed by K. Hiebert and L. F. Shampine [142]. The 'Illinois algorithm' [86] is used for actually locating the root.

In general, a vector of functions $g_i(t, y)$, $i = 1, 2, \ldots, \text{NG}$, may be supplied to DASSLRT such that the root of any of the NG functions g_i is desired. Of course, there may be several such roots in a given output interval. DASSLRT returns them one at a time, in the order in which they occur along the solution. An integer array tells the user which g_i, if any, were found to have a root on any given return. We note that because DASSLRT detects roots by looking for sign changes in g, it is conceivable for it to skip intervals where there are multiple roots. We have not found this to be a problem in the applications that we have seen.

At the time of this writing, DASSLRT exists and is routinely used to solve application problems at Lawrence Livermore National Laboratory and at Sandia National Laboratories, Livermore. A version for outside release is planned.

A second extension of DASSL has been developed to handle sensitivity analysis for DAE's. Suppose we are given a DAE

$$F(t, y, y', p) = 0, \qquad (5.3.2a)$$
$$y(t_0, p) = y_0(p) \qquad (5.3.2b)$$

5.3. OBTAINING NUMERICAL SOLUTIONS

where $y, y' \in \mathcal{R}^N$, whose solution depends on a vector of M time independent parameters p. We would like to compute the (N, M) matrix $W(t)$ of sensitivity functions

$$W(t) = \frac{\partial y(t)}{\partial p} \qquad (5.3.3)$$

which describes how the solution components change as a result of changes in the parameters.

The sensitivity matrix satisfies a system of DAE's which can be derived by partial differentiation of equations (5.3.2) with respect to the parameter vector p

$$\begin{aligned} \frac{\partial F}{\partial y'} W'(t) + \frac{\partial F}{\partial y} W(t) + \frac{\partial F}{\partial t} + \frac{\partial F}{\partial p} &= 0 \\ W(t_0) &= \frac{\partial y_0}{\partial p}. \end{aligned} \qquad (5.3.4)$$

Computationally, the important observation with respect to the sensitivity equations (5.3.4) is that they are linear and that they have the same iteration matrix as the original system (5.3.2) if they are solved using the same sequence of methods and stepsizes. Caracotsios and Stewart [66] have written a code DASSAC for solving the sensitivity system which makes use of these observations and is based on DASSL. The code works as follows. The sensitivity equations (5.3.4) are appended to the original system. The system is solved using a version of DASSL which has been modified to evaluate the iteration matrix on every step, require only one Newton iteration for the linear system (5.3.4), and make use of the repetitive block diagonal structure of the iteration matrix to save storage and operations in forming the matrix and solving the linear system. The error estimates and stepsize adjustment are based on the entire system (5.3.4) appended to (5.3.2). It is to be noted that the sensitivities may not be accurate if the iteration matrix is computed via finite differencing.

One of the obvious deficiencies of the standard DASSL is the lack of options for dealing with Jacobian matrices which do not fit naturally into a dense or banded category. This is particularly apparent in the case of large-scale problems from chemical engineering applications (see for example [110]) and in the method of lines solution of partial differential equations in more than one dimension. Marquardt [180] has recently implemented a version of DASSL which makes use of the HARWELL sparse direct linear system solver. A version of DASSL which uses the preconditioned GMRES method to solve the linear system at each Newton iteration is currently under development by Brown, Hindmarsh and Petzold, and is expected to be released soon. This code implements many of the strategies described for ODE's in [36]. Some initial experiences with the code are described in [204].

Finally, we note that Skjellum et al.[232] have recently implemented a version of DASSL for use on massively parallel computers, called Concurrent DASSL.

5.3.4 Determining Consistent Initial Conditions

Often the most difficult part of solving a DAE system in applications is to determine a consistent set of initial conditions with which to start the computation. More precisely, we formulate this problem as follows. Given information about the initial state of the system which is sufficient, in a mathematical sense, to specify a unique solution to the DAE, determine the complete initial state $(y(t_0), y'(t_0))$ corresponding to this unique solution. For example, the user may specify information about the solution and/or its derivatives at the initial time, and the problem is to determine the remaining values of the solution and its derivatives. In this subsection we describe a newly developed general algorithm for finding consistent initial conditions for index one and semi-explicit index two DAE systems. This work is described in more detail in Leimkuhler et al.[162].

For a simple algorithm, one might consider substituting $y = y_0$, $y' = y'_0$ and solving the resulting system, together with the user defined initial data, for y_0 and y'_0. For example, for an ODE $y' = Ay$ together with user defined information $Cy'_0 + Dy_0 = q_0$ at the initial point, we have

$$\begin{bmatrix} -A & I \\ D & C \end{bmatrix} \begin{bmatrix} y_0 \\ y'_0 \end{bmatrix} = \begin{bmatrix} 0 \\ q_0 \end{bmatrix}$$

which we can solve uniquely for y_0 and y'_0 provided the left hand matrix is of full rank and q_0 is in the range of the matrix $CA + D$. In this case, these are the same conditions which the matrices D and C must satisfy to ensure that the user has given enough information about the initial state to specify a unique solution to the mathematical problem.

On the other hand, consider the index one DAE given by

$$\begin{aligned} Y'_1 + Y'_2 + Y_1 &= g_1(t) \\ Y_2 &= g_2(t). \end{aligned} \quad (5.3.5)$$

To specify a unique solution to the DAE, it is sufficient to give the value of either Y_1 or Y'_1 at t_0, because Y_2 and Y'_2 must satisfy the constraint and the derivative of the constraint. However, evaluating Y_1, Y'_1, Y_2, and Y'_2 at time t_0 in (5.3.5), it is apparent that it is not possible to obtain $Y_1(t_0)$ uniquely if only $Y'_1(t_0)$ is specified since $Y'_2(t_0)$ is unknown. Thus this simple algorithm fails to give the solution, although the user has given enough information to specify a unique solution to the original problem.

Clearly the difficulty with the above procedure for equations (5.3.5) is that the simple algorithm has no way of obtaining the information about the derivative of the constraint which is inherent in the system. In the fully-implicit index one problem it is not in general possible to isolate the constraints and differentiate them. Thus we are led to consider the following algorithm, motivated by the definition of global index given in Chapter 2. We assume that the DAE system (5.3.1) has index ν.

5.3. OBTAINING NUMERICAL SOLUTIONS

Solve

$$F(t_0, y_0, y_0') = 0$$
$$\frac{dF}{dt}(t_0, y_0, y_0', y_0'') = 0 \qquad (5.3.6)$$
$$\vdots$$
$$\frac{d^\nu F}{dt^\nu}(t_0, y_0, y_0', \ldots, y_0^{(\nu+1)}) = 0$$

coupled with the user defined information

$$B(t_0, y_0, y_0') = 0. \qquad (5.3.7)$$

It is easy to show that if the user defined information (5.3.7) is sufficient to determine a unique solution to the DAE, then (5.3.6), (5.3.7) have a solution $(y_0, y_0', \ldots, y_0^{(\nu+1)})$, in which the first two components y_0, y_0' are uniquely determined and are the solution to the DAE at t_0. Note that the higher derivatives of y are not determined uniquely by this algorithm.

Since it is frequently not practical to obtain the derivatives of F analytically, we are led to consider approximating the derivatives of F. Because F is possibly only defined for $t \geq t_0$, it is natural to consider using one sided differences. For example, the simplest one sided difference is given by

$$D_h F = \frac{F(t_0 + h, y_0 + hy_0', y_0' + hy_0'') - F(t_0, y_0, y_0')}{h}.$$

We can define higher order difference approximations by

$$D_h F = \frac{1}{h} \left[\sum_{i=1}^{s} \alpha_i F(t_0 + hc_i, y_0 + hc_i y_0', y_0' + hc_i y_0'') - \left(\sum_{i=1}^{s} \alpha_i \right) F(t_0, y_0, y_0') \right] \qquad (5.3.8)$$

by choosing the constants α_i, c_i appropriately. The higher derivatives of F can be approximated similarly.

The algorithm obtained by replacing the consistency equations (5.3.6) by their approximations (5.3.8), coupled with the user defined information (5.3.7), produces a rank deficient over-determined nonlinear system. Unlike the analytic consistency equations, the approximate system may not have an exact solution because some of the derivative approximations to F may be approximating user defined information. Thus the approximate system is solved in a least squares sense. However, the solution of this system is complicated by the rank deficiency of the Jacobian. Using the structure of the system, it is possible to show that the minimum norm solution to this nonlinear least squares problem converges to the correct solution as the approximations D_h become more accurate. One can also formulate a scheme for replacing the system with a full rank system which has the same solution for y_0, y_0'. Numerical results for this technique appear to be promising.

In other approaches to the consistent initialization problem for DAE's, Berzins, Dew and Furzeland [16], in their code SPRINT, use the initialization method in DASSL to approximate the derivatives, and then improve these by also attempting to satisfy a difference approximation to the first time derivative of the DAE. The algorithm that we have described in this section can be considered to be a generalization of this idea. Pantelides [199] uses a graph theoretic algorithm to determine the minimal set of equations to be differentiated in order to solve for the consistent initial values. Mrziglod [190] gives an algorithm which is based on a thorough decomposition of the system structure. These latter two algorithms require knowledge of analytical expressions for various derivatives of the problem, which is a drawback for large scientific problems.

To our knowledge, no general purpose software for solving the consistent initialization problem is available at this time.

5.4 Solving Higher Index Systems

5.4.1 Alternate Forms for DAE Systems

In this subsection we consider some techniques for rewriting a DAE system in an alternative form that may be easier to solve numerically. All of the different forms of the equations that we consider are equivalent in the sense that, given a consistent set of initial conditions, different forms of a system have the same analytical solution. Computationally, however, some forms of the equations may have much different properties than others. We discuss some of the advantages and disadvantages of rewriting a high index DAE system in a different form.

As an example, let us consider some different ways to solve the constrained mechanical system given by

$$\begin{aligned} M(q)q'' &= f(q,q',t) + G(q)\lambda, \\ \Phi(q) &= 0. \end{aligned} \tag{5.4.1}$$

Here, the mass matrix M is nonsingular almost everywhere, λ is the Lagrange multiplier vector, and $\partial \Phi / \partial q = G^T$.

We can attempt to solve the system in its original, index three form, using an implicit numerical method such as BDF. This technique is actually used in some codes [197] for solving mechanical systems. Solving the problem in this way has the advantages that it is easy to formulate the system, as we do not have to differentiate the constraints or rewrite the system in any way, the sparsity of the system is preserved, and the constraints are satisfied exactly on every step. The disadvantage is that there are several difficulties in using a variable stepsize BDF code for solving systems in this form, which we describe in detail in Section 5.4.2. Many of the difficulties can be circumvented, but in general it is not a simple matter to obtain an accurate numerical solution for

5.4. SOLVING HIGHER INDEX SYSTEMS

the Lagrange multipliers λ. For BDF methods, λ and q' must be filtered out of the error estimate. A code based on this strategy does not always give reliable results for the variables which are filtered out of the estimate. In particular, results are likely to be unreliable in situations where there are steep gradients or discontinuities in the velocities. This type of error control strategy can also give incorrect solutions when the initial conditions are improperly specified in the sense that the algebraic constraint is satisfied but the derivative of the constraint is not. It may be possible to reliably solve systems in the index three form with methods other than BDF, such as extrapolation or defect correction, by controlling errors on the velocities as well as the positions, but we are not aware of any further work in this area.

A second way of solving (5.4.1) is to differentiate the constraint one or more times and solve the resulting lower index system. Except for the index zero system, there are still numerical difficulties, but they are less severe for the lower index systems. It should be noted that this type of lower index formulation of a problem does not force the constraints to be satisfied on every step, and there may be a tendency for the amount by which the constraint is not satisfied to increase from step to step. In Chapter 6 the 'drift' in the algebraic constraints is demonstrated in numerical experiments with DASSL on an index three pendulum problem when it is reformulated as an index one problem. By using small stepsizes, or in an automatic code by keeping the error tolerances fairly stringent, we can usually keep small the amount by which the constraint is not satisfied. Whether this is a serious problem or not depends on the application, although clearly it could be troublesome if the solution is desired over a long interval in time. In a recent paper, Führer and Leimkuhler [108] have shown by example that differentiating a nonlinear constraint can be dangerous in the sense that the stability properties of the new system will not be the same as the stability properties of the original system. Thus this approach should be used only with caution.

Another alternative is to use the index reduction techniques of Gear [118], which were described in Section 2.5.3, (see also Section 6.2) to rewrite the system as a lower index system. This approach requires differentiating the constraints and adding additional (algebraic) variables which act as Lagrange multipliers. This process can be performed repeatedly, to yield a semi-explicit index two system. The solution of this system satisfies both the constraints and the derivatives of the constraints. Because the solutions for the algebraic variables of this system are zero, the resulting system can be solved by quite general multistep methods, as described in Gear [117]. Another possibility for the constrained mechanical systems is to do the differentiation only once, and add the extra Lagrange multipliers. While the resulting system is a simpler index two system, not all of the algebraic variables are zero now. Therefore BDF methods, or appropriate implicit Runge-Kutta methods, can be used to solve the system. In the next section this promising approach using the BDF methods is pursued further. Another possibility suggested by Gear [118] for reducing the index in semi-explicit systems $x' = f(x, y, t)$, $0 = g(x, y, t)$, is to replace the undifferentiated variables y by w'. While the resulting system in x, w, t is one index lower, the solution of this new system yields only x and w so it is still necessary to differentiate the computed w to obtain y.

Yet another strategy, which is dependent upon knowing more information about the structure of the problem, and is used in some codes for solving mechanical systems (see Section 6.2), is to eliminate the Lagrange multipliers analytically using methods in analytical mechanics to obtain a standard form ODE system. The system of ODE's is assembled from a data structure describing the mechanical system. If the resulting problem is not stiff, this approach has the advantage that the system can be solved by an explicit numerical method. The number of unknowns after this type of transformation usually is smaller, but the sparsity of the system has decreased, which is an important consideration if the problem happens to be stiff. Again, we must be very careful that the initial conditions satisfy both the constraint and its derivative, or we will obviously obtain a solution which is nonsense. A constraint corresponding to an eliminated Lagrange multiplier is automatically satisfied in the chosen representation of the mechanical system. As an example, consider solving the equations of a pendulum in Cartesian coordinates. The system is written as

$$\begin{aligned} x'' &= \lambda x, \\ y'' &= \lambda y - g, \\ 0 &= \frac{1}{2}(x^2 + y^2 - L^2), \end{aligned} \quad (5.4.2)$$

where g is the gravity constant and L is the length of the bar. Let

$$x = L \sin \phi, \qquad y = -L \cos \phi.$$

Then the algebraic constraint of constant length is fulfilled and the well-known ODE is

$$\phi'' - \frac{g}{L} \sin \phi = 0.$$

This approach may be difficult to implement in general or for very large systems. Finally, we note that Führer and Leimkuhler [108] have recently studied the implications of the various formulations for mechanical systems. See also Section 6.2 for a further discussion on the formulation and solution of systems describing constrained mechanical systems.

A different way of dealing with the high index systems is through the use of penalty functions or regularizations. We loosely define the *regularization* of a DAE to be the introduction of a small parameter into the DAE in such a way that the solution of the perturbed system approaches the solution of the unperturbed system as the parameter tends to zero. Baumgarte [14] discusses a technique for circumventing the problem of 'drifting off' the constraints $\Phi(q)$ by adding to the original equations an equation consisting of a linear

5.4. SOLVING HIGHER INDEX SYSTEMS

combination of Φ, $d\Phi/dt$ and $d^2\Phi/dt^2$. The linear combination is chosen so that the resulting system damps errors in satisfying the constraint equation. This idea was used successfully by Adjerid and Flaherty [1] to stabilize the constraint in the solution of partial differential equations by adaptive moving mesh methods. The stabilization approach is similar, but not identical, to penalty function methods (Lötstedt [165], Sani et al. [219]). Depending on the choice of the parameters in the linear combination, we may see any of the difficulties discussed earlier. This technique introduces extraneous eigenvalues into the system, which may or may not cause difficulties. Finally, the penalty techniques have the disadvantage that if the initial conditions are not posed correctly, they introduce a nonphysical transient into the problem [219]. März [183] gives a regularization for the semi-explicit index two system so that the regularized system is index one. In recent experiments, Knorrenschild [153] reports that DASSL solves some regularized semi-explicit index two systems much more efficiently than systems which are not regularized. The regularization which Knorrenschild uses is motivated by physical considerations from circuit analysis applications but can be applied to general problems. We feel that at this time the results on regularization as a general technique for solving higher index systems are inconclusive. However, approaches based on some form of regularization have definitely proven successful in some applications. Regularization is discussed in more detail for multistep methods in Section 3.3, and for Runge-Kutta methods in Section 4.5.

Before leaving the subject of alternate forms for DAE systems, there is one more aspect of this problem that we wish to consider. Sometimes there is a choice of which variables to use for solving a problem. For example, in the system

$$\begin{aligned} u' &= v, \\ v' &= f(u,v,t) + G(u)\lambda, \\ G^T v &= 0 \end{aligned} \quad (5.4.3)$$

we could have replaced v in the constraint by u' to obtain

$$\begin{aligned} u' &= v, \\ v' &= f(u,v,t) + G(u)\lambda, \\ G^T u' &= 0. \end{aligned} \quad (5.4.4)$$

In implementing these systems, the question naturally arises, are there any advantages in writing the system in one form over the other? Using BDF methods combined with Newton iteration for linear problems in exact arithmetic, the two forms of the equations give identical solutions. This is because Newton's method is exact in one iteration for linear systems, and because the equations which result from discretizing both systems by BDF are identical. This last fact is obviously true for nonlinear systems too. For nonlinear systems in exact arithmetic we know of no reasons why Newton's method would be more likely to converge for one form of the equations than the other. Both forms of the system may lead to very poorly conditioned iteration matrices. In the next section we will suggest scaling techniques to overcome this difficulty. These techniques are directly applicable to systems such as (5.4.3) which are in Hessenberg form, but not to systems such as (5.4.4) which are not in any standard recognizable form. Equation (5.4.3) may be preferred for just this reason. This question of which variables to use in writing a system may seem like a minor point, but in adaptive moving mesh calculations for partial differential equations we found that the formulation of the problem in this sense was crucial [203]. Sometimes, as happened in this application, a variation in the problem formulation which is equivalent in exact arithmetic to the original formulation can drastically change the condition of the iteration matrix and dramatically affect the reliability and efficiency of the calculation. We will investigate this adaptive moving mesh example further in Chapter 6. In general, we recommend investing some time into the problem formulation step of the solution process, keeping in mind the standard forms which have so far proven so the most successful.

5.4.2 Solving Higher Index Systems Directly

Although convergence rates are known for BDF and implicit Runge-Kutta methods applied to some forms of semi-explicit higher index systems, the implementation of codes based on these methods for solving higher index systems is not straightforward. Here we examine some of the difficulties and suggest some remedies.

Matrix ill conditioning is a problem for numerical methods when applied to DAE systems, and especially to higher index systems. We specifically study the conditioning of the iteration matrices corresponding to the BDF methods and useful (i.e., efficiently implementable) IRK methods, such as the SIRK's or DIRK's. It is not practical to consider general IRK methods where the stage equations cannot be uncoupled, either naturally or via transformations in the case of SIRK's, due to the high cost of solving large nonlinear systems. Even so, the iteration matrices for general IRK methods will contain blocks with a structure resembling the one studied below. Hence their conditioning is expected to be very similar. From Theorem 2.3.4 we have

Theorem 5.4.1 *The condition number of the iteration matrix for a system with (local) index of ν is $O(h^{-\nu})$.*

It is often possible to reduce the condition number by scaling the DAE system appropriately. This is especially true for semi-explicit systems

$$\begin{aligned} x' &= f(x,y,t) \\ 0 &= g(x,y,t). \end{aligned}$$

5.4. SOLVING HIGHER INDEX SYSTEMS

For these systems, the iteration matrix is written as

$$hJ_n = \begin{bmatrix} \alpha_0 I - h\frac{\partial f}{\partial x} & -h\frac{\partial f}{\partial y} \\ h\frac{\partial g}{\partial x} & h\frac{\partial g}{\partial y} \end{bmatrix}.$$

We consider two cases: semi-explicit index one systems, and semi-explicit index two systems with the property that $\partial g/\partial y \equiv 0$. Scaling for index three systems arising in constrained mechanical problems is discussed in [205].

Case I. When the index is one, we have that $\partial g/\partial y$ is nonsingular, so hJ_n is nonsingular as $h \to 0$ if we scale the rows corresponding to the algebraic constraints by $1/h$. Since we are not scaling variables, but only equations, the effect of this scaling should be to improve the accuracy of the solution of the linear system, for *all* variables.

Case II. For this case, we will assume that the index is two, and that $\partial g/\partial y \equiv 0$. By explicitly computing $(hJ_n)^{-1}$ we find that the orders of the blocks of the inverse are

$$\begin{bmatrix} 1 & 1/h \\ 1/h & 1/h^2 \end{bmatrix},$$

where the elements in the first row correspond to x and those in the second row to y. Roundoff errors proportional to u/h and u/h^2 are introduced in x and y, respectively, while solving this unscaled linear system. See Lötstedt and Petzold [205] for a more detailed explanation. If we scale the bottom rows of hJ_n (corresponding to the 'algebraic' constraints) by $1/h$, then the scaled matrix can be written as

$$h\hat{J}_n = \begin{bmatrix} \alpha_0 I - h\frac{\partial f}{\partial x} & -h\frac{\partial f}{\partial y} \\ \frac{\partial g}{\partial x} & 0 \end{bmatrix}$$

With this scaling the roundoff errors are $O(u)$ in x and $O(u/h)$ in y. As $h \to 0$ these errors can begin to dominate the solution y. In this case, the error estimates and convergence tests in an automatic code may experience difficulties due to these growing rounding errors. These difficulties, along with what can be done to minimize their effects, will be described in greater detail later. For now, we merely note that the effect of the proposed scaling is to control the size of the roundoff errors in x which are introduced in solving the linear system. At the same time, the algebraic variables y may contain errors proportional to u/h. However, since the values of y do not affect the state of the system directly (that is, how the system will respond at future times), we may be willing to tolerate much larger errors in y than in x. In any case, this scaling is a significant improvement over the original scaling. For the scaled system, the errors are considerably diminished, and the largest errors are confined to the variables which are in some sense the least important. Painter [198] describes difficulties due to ill conditioning for solving incompressible Navier-Stokes equations and employs essentially the same scaling that we have suggested here to solve the problem. These difficulties are most severe when an automatic code is using a very small stepsize, as in starting a problem or passing over a discontinuity in some derivative.

We further note that if we are using Gaussian elimination with partial pivoting, we do not need any column scaling. The solution will be the same without this scaling, because it does not affect the choice of pivots [235]. What the analysis shows is that the errors which are due to ill conditioning are concentrated in the algebraic variables y of the system and not in the differential variables x. Thus, we must be particularly careful about using the "algebraic" variables in other tests in the code which might be sensitive to the errors in these variables.

There is a simple way to implement row scaling in a general DAE code, which unfortunately requires a minor change to DASSL. By modifying the argument list for the RES routine to include the stepsize h, the user can scale DELTA inside this subroutine. This way, the scaling costs virtually nothing. An alternative idea is to provide an option to automatically do row equilibration, or to use linear system solvers which perform row scaling, as suggested by Shampine [222] for stiff ODE systems.

Even after scaling, for high index systems there are relatively large errors in some of the algebraic variables. Since these variables do not determine the future state of the system, the errors are in some sense tolerable. For an automatic code we must have some criterion for deciding when to terminate the corrector iteration. It is shown in Petzold and Lötstedt [205] that to maintain the global order of accuracy of the state variables of the system, it is sufficient to terminate the Newton iteration based on the errors of the *scaled* variables for the column scaling which would bring the condition number of the system to $O(1)$. For the index two system where $\partial g/\partial y \equiv 0$, the iteration error of hy, rather than y, would be required to be less than a prescribed tolerance. If y needs to be known very accurately, this strategy is not appropriate.

It remains to examine the integration error tests which are used to control the stepsize in an automatic code. For index one systems and for index two systems where $\partial g/\partial y \equiv 0$, the errors in the algebraic variable y on previous time steps do not directly influence the errors in any of the variables at the current time [205]. Therefore, we can consider deleting these variables from the error estimate in order to promote the smooth operation of a code. Consider first the case of solving semi-explicit index one systems. Suppose, for example, that on one step we make a fairly large mistake by terminating the Newton iteration before it has really converged, and that on this step the value of y that should have been computed has a large error in it, but that based on the incorrect value it passes the error test anyway. If we base the error test on y, then a bigger problem may arise on the next step because the new value of y does not approach the old (incorrect) value of y, so that their difference, and hence the error estimate, does not approach zero as $h \to 0$. For this reason we feel that it is probably wise to leave y out of the error control decisions in semi explicit index one systems. The main drawback in this strategy is that if

5.4. SOLVING HIGHER INDEX SYSTEMS

we need to know the values of y at interpolated points (between mesh points), then the stepsize should be based in part on the values of y. For index two systems, the stepsize control strategy in an automatic BDF code will fail, for reasons explained in Petzold [200], unless the algebraic variables y are excluded from the integration error test.

Another possibility for BDF codes is to construct an error estimate which automatically 'filters' errors corresponding to the algebraic variables out of the estimate, an idea introduced in [179] and [229] for linear constant coefficient systems of arbitrary index ν. This filtering approach has been studied further for more general index one and two systems in [160,205]. A similar strategy is applied to control the errors in the solution of stiff problems by Sacks-Davis [217,224].

To motivate the estimate of interest here for index one and semi-explicit index two systems, consider applying the implicit Euler method to the system

$$Ax' + Bx = f(t). \qquad (5.4.5)$$

Taking one step, we obtain

$$A\left(\frac{x_{n+1} - x_n}{h_{n+1}}\right) + Bx_{n+1} = f(t_{n+1}). \qquad (5.4.6)$$

The true solution to (5.4.5) satisfies

$$A\left(\frac{x(t_{n+1}) - x(t_n) - \frac{h_{n+1}^2}{2}x''(\xi)}{h_{n+1}}\right) + Bx(t_{n+1}) = f(t_{n+1}). \qquad (5.4.7)$$

Subtracting (5.4.7) from (5.4.6), we obtain

$$A\left(\frac{e_{n+1} - e_n + \frac{h_{n+1}^2}{2}x''(\xi)}{h_{n+1}}\right) + Be_{n+1} = 0, \qquad (5.4.8)$$

where $e_{n+1} = x_{n+1} - x(t_{n+1})$. Solving for e_{n+1}, we have

$$e_{n+1} = (A + h_{n+1}B)^{-1}Ae_n - (A + h_{n+1}B)^{-1}A\left(\frac{h_{n+1}^2}{2}\right)x''(\xi).$$

Thus the contribution to the global error from the current step is given by

$$e_{local} = (A + h_{n+1}B)^{-1}A\tau_{\ell te}, \qquad (5.4.9)$$

where $\tau_{\ell te}$ is the local truncation error in the current step. A similar argument for higher order BDF gives (5.4.9) as the contribution to the global error from the current step, where $(A + h_{n+1}B)$ has been replaced by $(A + h_{n+1}\alpha_0 B)$. Here, α_0 is a constant depending on the method. The application of this filtering idea to other methods is not advised without an error expression to justify it. It can cause asymptotically incorrect estimates if it is carelessly used in combination with other types of methods and local truncation error estimates.

It is useful to reflect on the effects of the filtering for some systems of interest. For index one systems, the filter has the effect of removing the components of the local truncation error corresponding to the algebraic variables from the error estimate because multiplying by the matrix A kills these components. For implicit ODE systems (A nonsingular), the filter tends to the identity as $h_{n+1} \to 0$, giving the usual estimate. For semi-explicit index two systems with $\partial g/\partial y \equiv 0$, the filtering yields an estimate for the kth order BDF which is $O(h^{k+1})$ for the differential variables and $O(h^k)$ for the algebraic variables, thereby correctly reflecting the behavior of the local error. On DAE's with index greater than two, it is quite likely that DASSL (or any other BDF code for DAE's) will fail even when this filter is applied. A stronger filter or a different strategy altogether may be required to control the errors in these high index DAE's. In [205] Petzold and Lötstedt suggest a strategy for error control of index three mechanical systems.

Finally, the estimate (5.4.9) is easily computed in a code like DASSL. The matrix $(A + h_{n+1}B)$ is the iteration matrix used in the solution of the corrector equation. Hence it is already available in factored form. The term $A\tau_{\ell te}$ can be approximated by noting that for $F(t, y, y') = 0$, $A = \partial F/\partial y'$ and $\tau_{\ell te}$ is the usual local truncation error estimate. Then $(\partial F/\partial y')\tau_{\ell te}$ can be approximated by

$$\frac{\partial F}{\partial y'}\tau_{\ell te} \equiv F(t, y, y' + \tau_{\ell te}) - F(t, y, y').$$

Even more simply for semi-explicit systems, the quantity $A\tau_{\ell te}$ is a vector composed of the local truncation errors corresponding to the differential variables x and zeros corresponding to the algebraic variables y. The filter is used in the decision on whether to accept or reject the step. The order selection remains unchanged as it is based on a comparison of relative magnitudes of terms in a Taylor series, rather than on relative magnitudes of local errors. Berzins [17] has recently reported good success in applying part of the filter (namely, the matrix A) also in the order selection process. This approach has the disadvantage that the order selection is not then independent of scalings of the system. This latter point is not an issue for semi-explicit systems.

The filtered estimate is also advantageous for stiff systems which experience frequent steep transients of small magnitude. For example, in the solution of partial differential equations by operator splitting techniques or with adaptive mesh methods, there is a discontinuity and a steep transient after every rezone of the mesh. The filtered estimate allows the code to take relatively large steps through the transient, which would not be allowed using the usual estimate. The filtered estimate has been shown to be quite effective [16,204] in reducing the number of time steps without sacrificing accuracy in experiments with adaptive grid solutions of partial differential equations.

Chapter 6

Applications

6.1 Introduction

In Chapter 1 we briefly discussed a number of scientific disciplines in which DAE's have occurred. Now we consider some of these areas in more depth in order to discuss several issues which are critical to the numerical solution process. For example, determining consistent initial values for a DAE system often requires special problem-dependent techniques. There are often alternative formulations of the DAE for a particular application. The performance of a numerical method can be influenced dramatically by the choice of problem formulation. Our focus in this chapter is to examine these issues as they pertain to specific applications, rather than to give a comprehensive and general description of all the various techniques. Examples arising in the simulation of mechanical systems, electrical systems, trajectory prescribed path control problems, and the discretization of partial differential equations by the method of lines (MOL) will be used to illustrate these ideas and techniques.

Historically, DAE's have been solved indirectly by requiring the system to be reformulated as an explicit ODE system, for which there are many software packages. Often, the reduction of a DAE to an ODE system requires complicated problem-specific manipulations of the equations, incorporation of boundary conditions into the system equations and simplification of the underlying problem. This process slows the study of more complex systems because it often has to be repeated when the model is altered. The ease of setting up and altering DAE models is a major factor in the large amount of current interest in them.

There are often several natural ways to formulate a given application. Different formulations may affect the behavior of the numerical solution process. Sometimes there are choices in both the selection of the equations to be imposed and the variables. In many cases, there is the option of reducing the index of the system being considered. But it is not always apparent whether the reduced index formulation of a problem will yield a better implementation. If the numerical solution is obtained by solving a reduced index problem, the satisfaction of the algebraic equations in the original system may gradually deteriorate as the integration process continues. This deterioration occurs because most index reduction techniques produce a reduced index DAE where the original constraints are no longer directly imposed. However, it is possible, by adding some extra variables and equations as described in Section 2.5 [121], to derive a reduced index system where the original constraints are still imposed. The problem formulation may also affect the conditioning and the sparsity structure of the linear systems which must be solved at every time step for a DAE. We have discussed some of these general issues in detail in earlier chapters. Here, we examine them from the point of view of the solution of DAE's arising in applications from science and engineering.

In Section 6.2 we outline some of the issues involved in the solution of constrained mechanical systems, and give numerical results illustrating the performance of numerical methods for different formulations of the problems. In Section 6.3 we examine the solution of DAE systems arising in the trajectory prescribed path control of space vehicles. These are severely nonlinear index three systems which are in Hessenberg form. We compare several different formulations and numerical methods applied to these problems. One of the earliest and most successful applications of numerical methods for solving DAE systems directly was the modeling of electrical networks, which we discuss in Section 6.4. Usually these problems yield index one systems, although it is possible to design circuits which give rise to arbitrarily high index DAE's. Another area where DAE's have recently been playing an increasingly important role is in the solution of partial differential equations via the method of lines. We discuss applications from combustion ignition, incompressible fluid flow, adaptive moving meshes, and the modeling of gas transmission networks in Section 6.5, examining issues such as problem formulation, determination of consistent initial conditions, and structure and conditioning of linear systems arising in the solution process.

6.2 Systems of Rigid Bodies

The motion of systems of rigid bodies can be described using principles of classical mechanics as the solution of a constrained variational problem [4,189]. The corresponding Euler equations of this class of problems is a DAE, which is called in classical mechanics the Lagrange equation of the first kind. This DAE has the following structure

$$M(x)x'' = g(x, x', t) + G^T(x)\lambda \quad (6.2.1a)$$
$$0 = \phi(x), \quad (6.2.1b)$$

with $G = \phi_x$, $x \in R^n$, $\lambda \in R^m$, and $m \leq n$.

For example, in a constrained mechanical system with position x, velocity u, kinetic energy $T(x, u)$, external force $f(x, u, t)$, and constraint $\phi(x) = 0$,

6.2. SYSTEMS OF RIGID BODIES

the Euler equation is

$$x' = u \tag{6.2.2a}$$
$$\frac{d}{dt}\frac{\partial}{\partial u}T(x,u) = \frac{\partial T}{\partial x} + f(x,u,t) + G^T\lambda \tag{6.2.2b}$$
$$0 = \phi(x), \tag{6.2.2c}$$

where $G = \partial\phi/\partial x$ and λ is the Lagrange multiplier. This system can be rewritten as

$$x' = u \tag{6.2.3a}$$
$$T_{uu}u' = g(x,u,t) + G^T\lambda \tag{6.2.3b}$$
$$0 = \phi(x), \tag{6.2.3c}$$

where $T_{uu} = \partial^2 T/\partial u^2$. This DAE system in the unknown variables u, x and λ is linear in the derivative. If, as is often the case, T_{uu} is a positive definite matrix, then multiplication of (6.2.3b) by T_{uu}^{-1} converts (6.2.3) to a semi-explicit index three DAE. The equations (6.2.3) are often called the *descriptor form* of the equations of motion. Although most of these systems are index three, when the mechanical systems are modeled together with control systems, as often occurs in the simulation of robots, the resulting systems can have arbitrarily high index [89,90,91].

A simple example of such a problem is the pendulum problem which has been discussed throughout this text. Recall that the equations modeling the motion, in Cartesian coordinates, of a simple pendulum of length L, assuming an infinitesimal ball of mass one at one end, tension λ in the bar, and gravitational force g, are given by

$$x'' = -\lambda x \tag{6.2.4a}$$
$$y'' = -\lambda y - g \tag{6.2.4b}$$
$$0 = x^2 + y^2 - L^2. \tag{6.2.4c}$$

It is well known that this problem can be described in a more compact form by eliminating the constraints, which results in the ODE

$$\phi'' - \frac{g}{L}\sin\phi = 0, \tag{6.2.5}$$

where $x = L\sin\phi$ and $y = -L\cos\phi$. This form of the equations is known in system theory and mechanics as the *state space form* or Lagrange equation of the second kind. In general, the state space form of a DAE is an explicit system of ordinary differential equations in a minimal set of variables which completely describes the system. This representation of the system may have a significantly smaller dimension than the descriptor form (6.2.1), which may contain redundant information about the system.

While both forms of the equations for the pendulum example can be established easily using pencil and paper, for more complicated systems like mechanisms, robots, vehicles or spacecraft, this is no longer feasible. For this purpose, so-called multibody formalisms have recently been developed which enable these equations or computer programs evaluating these equations to be generated automatically [193,212]. The performance of computer programs to model these systems depends strongly on the choice of coordinates x. If absolute coordinates are taken [197], the resulting equations have the form (6.2.1), with extremely sparse matrices (M being block diagonal). The effort to generate these equations and to evaluate them is small. On the other hand, more sophisticated approaches using relative coordinates are able to establish the state space form directly for an important class of mechanical systems. directly. These systems are the so-called *tree-configured systems*. Tree-configured systems have the property that when the individual bodies of a multibody system are related to the vertices, and the joints to the edges of a graph, the resulting graph has a tree structure. This form can be numerically treated by standard ODE methods, and is of much lower dimensionality than the descriptor form. However, the corresponding matrices are normally dense. For more details on this approach, see [241].

For systems which are not tree-configured, there are loop-closure conditions which remain as algebraic equations and which cannot in general be solved analytically. For example, in the modeling of a robot, the system is tree-configured if the robot hand is not in contact with some surface (as this would generate a closed loop). For these systems, traditionally the attempt was made to establish the state space form numerically [241,193]. This is easy to do in the case of linear constraints [109], but in the general case the coordinates may change from integration step to integration step or even in between.

In multibody mechanics, the number of unknowns is usually between ten and several hundred. Thus it is generally much smaller than in some of the other disciplines to be discussed later in this chapter. Therefore, both the descriptor form and the more traditional state space form have been realized in the current multibody codes.

There are several possibilities for the formulation of the descriptor system (6.2.3). All of these possiblities have the same analytic solutions. However, numerical methods applied to these systems can often exhibit quite different behavior. One such system can be obtained by replacing the *position-level* constraint (6.2.3c) by its derivative with respect to time,

$$0 = G(x)u. \qquad (6.2.6)$$

This constraint is sometimes called the *velocity-level* constraint. The resulting system has index two. It has the advantage that the index is lower, and the disadvantage that the position-level constraint is no longer automatically forced to be satisfied, thus allowing the possibility of the solution to drift away from that constraint. The constraint (6.2.6) can be differentiated once again

6.2. SYSTEMS OF RIGID BODIES

to obtain the *acceleration-level* constraint

$$0 = G(x)u' + u^T G_x u. \qquad (6.2.7)$$

The resulting system has the apparent advantage of having index one, but suffers from the disadvantage that the solution may drift away from satisfying the position-level and velocity-level constraints. Equation (6.2.3b) can also be solved for u', and the result introduced into (6.2.7) yields yet another index one DAE.

The extent to which drift can be a problem is highly problem-dependent. However, Führer and Leimkuhler [108] have recently shown that in general, the differentiation of a *nonlinear* constraint can be dangerous in the sense that the stability properties of the resulting system are not the same as those of the state space form. The inadvertent creation of an unstable system through differentiation of the constraints can lead to a very inefficient, or in the worse case, inaccurate numerical method. Thus in general we recommend stabilizing the constraints by some means as described below.

Another index two formulation of this problem can be obtained by introducing new variables μ, replacing (6.2.3a) by

$$x' = u + G^T \mu, \qquad (6.2.8)$$

and appending the new algebraic equation $Gu = 0$ to the system. We will call the resulting system the *stabilized index two formulation* of the problem. The idea of reformulating the system in this way first came from Gear et al. [121], where it is shown that if T_{uu} is positive definite and if G has full rank, then any solution of the original index three system (6.2.3) is a solution of the stabilized index two system, and conversely. The advantage of this formulation is that the position-level and velocity-level constraints are automatically enforced, thus eliminating the problem of drift for these constraints. The extra variables are not really much of a disadvantage, because the computations can be arranged in a form that avoids the storage of the new variables and requires very little computation [108,121]. This reformulation of the equations is discussed in greater detail in Section 2.5.3.

Finally, in [107,211,234] it has been proposed to use both the original constraints and one or more derivatives of the constraints simultaneously in the DAE formulation. This approach results in an overdetermined system of DAE's consisting of (6.2.3) coupled with (6.2.6) and (6.2.7). This system has the same analytic solutions as (6.2.3). It would first appear that numerical methods based on this technique would not be advantageous computationally, because of the cost of the linear algebra for the solution of the overdetermined system. However, it was recently shown by Führer and Leimkuhler [108] that if the stabilized index two formulation is extended to maintain also the acceleration level (by means of more Lagrange multipliers), then the resulting methods give the same numerical solutions as the overdetermined system approach. Further techniques for improving the efficiency of this approach are given in [108].

Yet another technique for stabilizing the constraint equations was proposed by Baumgarte [14]. In this approach, the acceleration-level constraint is replaced by a linear combination of the acceleration-level, velocity-level and position-level constraints. The coefficients in the linear combination are chosen in such a way that any errors which result from drifting off the constraints are damped out rapidly. This approach is analogous to the use of penalty functions in solving optimization problems. It has the disadvantage of introducing extraneous eigenvalues into the system. For too-rapid damping, this can result in extra stiffness or other problems. We are not aware of any general methods for choosing the values of the coefficients. On the other hand, this approach is quite general and has proven to be useful in several applications even outside the area of mechanical systems [1].

Another way to introduce penalty functions has been discussed by Lötstedt [165]. This approach applies the penalty terms to the ODE of the corresponding unconstrained mechanical system. These terms can be physically interpreted as additional springs and dampers forcing the system to satisfy the constraints approximately. Lötstedt showed by applying techniques from singular pertubation theory that the solution of this approach, which is rather common in engineering practice, converges to the exact solution of the constrained problem if the stiffness and damping coefficients tend to infinity.

We will illustrate some of the implications of the problem formulation by means of numerical experiments with the pendulum problem. First, let us rewrite the system (6.2.4) as a first order system of ODE's coupled with an algebraic equation, by letting $z_1 = x$, $z_2 = y$, $z_3 = x'$, $z_4 = y'$

$$z_1' = z_3 \qquad (6.2.9a)$$
$$z_2' = z_4 \qquad (6.2.9b)$$
$$z_3' = -z_1 \lambda \qquad (6.2.9c)$$
$$z_4' = -z_2 \lambda - g \qquad (6.2.9d)$$
$$0 = z_1^2 + z_2^2 - L^2. \qquad (6.2.9e)$$

System (6.2.9) is an index three semi-explicit nonlinear DAE. By differentiating the algebraic constraint (6.2.9e) once, and substituting for the derivatives of the state variables z_1, z_2, one obtains a new algebraic constraint

$$z_1 z_3 + z_2 z_4 = 0. \qquad (6.2.10)$$

Equations (6.2.9a)-(6.2.9d) and (6.2.10) form an index two DAE in semi-explicit form. Differentiating the algebraic constraint (6.2.10) one more time, substituting for z_1', z_2', z_3', z_4', and using (6.2.9e), the corresponding index one DAE system includes the differential equations and the algebraic equation

$$z_3^2 + z_4^2 - \lambda L^2 - z_2 g = 0. \qquad (6.2.11)$$

To obtain the stabilized index two formulation for the pendulum problem, introduce a new variable μ, and replace the differential equations (6.2.9a) and

6.2. SYSTEMS OF RIGID BODIES

(6.2.9b) by

$$z_1' = z_3 + z_1\mu \quad (6.2.12a)$$
$$z_2' = z_4 + z_2\mu, \quad (6.2.12b)$$

yielding a new DAE system composed of equations (6.2.12), and (6.2.9c)-(6.2.10).

We now give numerical results for three different formulations of the pendulum problem. For our experiments, we take $L = 1$, $g = 1$, and consistent initial values $z_1(0) = 1$, $z_2(0) = 0$, $z_3(0) = 0$, $z_4(0) = 1$, and $\lambda(0) = 1$ at $t = 0$. A truth model for the solution at $t = 1$ was generated by solving the index one system by DASSL with a tight local error control (RTOL = 1.0E-12).

First, a simple code implementing the two-step BDF method with a constant stepsize was used to compute the numerical solution of the index three pendulum problem for a sequence of fixed stepsizes [31]. As a predictor, the code utilized the interpolating polynomial based on the last three functional values of the solution, so that two additional starting values were required. They were obtained from the truth model. The Newton iteration was terminated when the correction terms were sufficiently small. The relative errors in each variable at $t = 1$ are given in Table 6.2.1, where it can be seen that the rate of convergence remained second order in all the variables.

Variable	Solution	h = .01	h = .005	h = .0025
z_1	0.1349949261	2.6E-3	6.5E-4	1.6E-4
z_2	0.9908462897	4.8E-5	1.2E-5	3.0E-6
z_3	-1.710951582	2.7E-4	6.6E-5	1.7E-5
z_4	0.2331035448	2.5E-3	6.2E-4	1.6E-4
λ	3.972538869	1.9E-4	4.7E-5	1.2E-5

Table 6.2.1: Index Three Pendulum Problem
Relative Errors in Two-Step BDF Solution at $t = 1$

To show how the numerical solution of the index one formulation drifts away from the original constraint manifold, we solved the index one version of the pendulum problem, namely (6.2.9a)-(6.2.9d), and (6.2.11), by DASSL for a sequence of local error tolerance inputs. To measure the distance of the numerical solution from the constraints, the numerical solution at $t = 1$ was substituted into the constraints

$$\begin{aligned} G(1) &= 1 - z_1^2 - z_2^2 \\ G(2) &= z_1 z_3 + z_2 z_4 \\ G(3) &= z_3^2 + z_4^2 - \lambda - z_2. \end{aligned} \quad (6.2.13)$$

The values $G(i)$, $i = 1, 2, 3$ are summarized in Table 6.2.2 along with the total numbers of integration steps (STEPS) and evaluations (FNS) of the DAE that were required for each case. Observe that the original $G(1)$ constraint is satisfied to lower accuracy than RTOL. Similarly, for RTOL $\leq$ 1.E-9, there is some drift in satisfying the constraint $G(2)$. As expected, $G(3)$ does not exhibit any drift, and the errors for this constraint are of the same magnitude as the tolerance which terminates the Newton iteration on each step.

RTOL	G(3)	G(2)	G(1)	STEPS	FNS
1.E-5	.166E-5	.321E-4	-.363E-4	43	89
1.E-6	.805E-8	-.163E-6	.542E-5	53	114
1.E-7	-.238E-8	-.474E-7	-.134E-6	84	164
1.E-8	-.798E-8	-.484E-7	.119E-6	90	197
1.E-9	-.405E-11	-.153E-7	.251E-7	116	254
1.E-10	-.173E-10	-.236E-8	.286E-8	155	359
1.E-11	.128E-12	-.208E-9	.252E-9	233	524
1.E-12	-.284E-13	-.652E-11	.196E-10	369	642

Table 6.2.2: DASSL Solution of Index One Pendulum Problem Algebraic Constraint Residuals at $t = 1$

Finally, we give the results of solving the stabilized index two version of the pendulum problem, that is equations (6.2.9c)-(6.2.10), (6.2.12a), and (6.2.12b), by DASSL. Here, the algebraic variables have been excluded from the integration error control estimates. All the variables were included in the convergence test for the corrector iteration. The relative errors in the numerical solution are given in Table 6.2.4 for three different values of RTOL. Table 6.2.3 shows how well the algebraic constraints corresponding to the index two and three formulations are satisfied at $t = 1$. Note that this index two formulation forces the satisfaction of both algebraic constraints (6.2.9e) and (6.2.10) to the prescribed tolerances, thereby eliminating the problem of drift as seen earlier in Table 6.2.2. For tolerances of 1.E-8 and smaller, the corrector iteration was unable to converge because the corrector test was too stringent for the algebraic variables.

RTOL	G(1)	G(2)	STEPS	FNS
1.E-5	.288E-6	-.437E-6	35	101
1.E-6	.915E-7	-.840E-7	45	123
1.E-7	.643E-9	-.151E-8	80	225

Table 6.2.3: DASSL Solution of Index Two Pendulum Problem Algebraic Constraint Residuals at $t = 1$

6.3. TRAJECTORY PRESCRIBED PATH CONTROL

RTOL	z_1	z_2	z_3	z_4	λ	μ
1.E-5	2.8E-4	5.4E-6	2.2E-5	3.1E-4	8.6E-6	8.3E-5
1.E-6	2.3E-5	3.7E-7	1.4E-6	2.2E-5	4.2E-6	4.3E-6
1.E-7	2.8E-6	5.2E-8	6.0E-8	2.8E-6	1.3E-5	1.0E-5

Table 6.2.4: Index Two Pendulum Problem
Relative Errors in DASSL Solution at $t = 1$ (Absolute Errors in μ)

6.3 Trajectory Prescribed Path Control

In the simulation of space vehicles, the shape of the trajectory is often prescribed by appending a set of path constraints to the equations of motion. The model equations then become a nonlinear semi-explicit DAE system. The differential equations include the classical equations of motion, and the algebraic equations correspond to the path constraints. These types of problems are called *trajectory prescribed path control* (TPPC) problems. Historically, specific TPPC problems have been solved by developing software models which instantaneously correct the control variables in order to approximately follow the prescribed path profiles. Here we examine general numerical techniques which can be applied directly to the DAE's. TPPC problems occur in many areas including the robot control problems of the preceding section and chemical process control [199]. We limit the discussion in this section to space vehicles.

The DAE systems arising in TPPC simulations typically involve a set of ℓ_1 *state equations*, including the equations of motion

$$x' = f(t, x, u), \qquad (6.3.1)$$

and ℓ_2 algebraic *path constraints*

$$0 = g(t, x, u), \qquad (6.3.2)$$

where u has dimension ℓ_2 and $\ell_2 \leq \ell_1$. Traditionally, the x variables are referred to as *state variables*, while the u variables are called the *control* or *algebraic variables*. To be consistent with our earlier terminology, we shall refer in this section to the x variables as the *differential variables*. The differential variables describe the position, velocity, and sometimes the mass of the vehicle. The control variables are typically angle of attack α and/or bank angle β. When the Jacobian matrix g_u is nonsingular, the path equations are referred to as *control variable constraints* and the corresponding DAE has index one. Often the path constraints in a TPPC problem are functions only of the differential variables, so that $g_u \equiv 0$ and the DAE will be of higher index. The path equations are then said to be *state variable constraints*. Note that TPPC systems are semi-explicit DAE's and are nonlinear in both the differential and algebraic variables.

The differential variables are often expressed in an *inertial* coordinate system. For example, position and velocity of the vehicle are given as the standard rectangular three-component vectors $\vec{r}$ and $\vec{v}$, respectively. The equations of motion can be written as

$$\vec{r}\,' = \vec{v}$$
$$\vec{v}\,' = \vec{g}(\vec{r}) + \frac{1}{m}\left(\vec{F}_A(\vec{r},\vec{v},\alpha,\beta) + \vec{F}_P(\vec{r},\alpha,\beta)\right) \qquad (6.3.3)$$

with $\vec{r}(t_0)$, $\vec{v}(t_0)$ and t_0 given, and

$$\begin{aligned}
\vec{g}(\vec{r}) &= \text{acceleration vector due to gravity,} \\
\vec{F}_A(\vec{r},\vec{v},\alpha,\beta) &= \text{aerodynamic force vector,} \\
\vec{F}_P(\vec{r},\alpha,\beta) &= \text{propulsive force vector,} \\
m &= \text{mass of the vehicle.}
\end{aligned}$$

There may also be other differential equations describing, for example, the rate of change of the vehicle's mass or temperature at selected surface points. This formulation of the differential equations enables a trajectory simulation program to be structured in terms of physical models. For example, separate software models are developed for the gravity, aerodynamic and propulsive forces. Then, for a particular application, the user selects the specific physical models the application requires. The trajectory program automatically defines the appropriate terms in the differential equations. Simulations of TPPC problems can easily be built, and later modified, to include the effects of the most realistic physical models for the various forces.

It is also useful to express the differential variables in a *relative* coordinate system since typically the path constraints are most naturally formulated using these coordinates. Figure 6.3.1 illustrates the equivalent spherical representation of the position and velocity variables: altitude H, longitude ξ, geocentric latitude λ, magnitude V_R of the relative velocity vector $\vec{R}^R$, relative flight path angle γ, and relative azimuth A. As shown in Figure 6.3.1, the position quantities H, ξ, and λ are measured with respect to the relative coordinate system (X_R, Y_R, Z_R), while the velocity variables V_R, γ, and A are measured in a local horizontal frame of reference (X_{LH}, Y_{LH}, Z_{LH}).

We limit our discussion to reentry vehicles with no propulsive forces so that $\vec{F}_P \equiv 0$, and simplify the simulation to model only a spherical geopotential and spherical earth. The equations of motion in relative coordinates are then given by

$$\begin{aligned}
H' &= V_R \sin(\gamma) \\
\xi' &= \frac{V_R \cos(\gamma) \sin(A)}{r \cos(\lambda)} \\
\lambda' &= \frac{V_R}{r} \cos(\gamma) \cos(A)
\end{aligned}$$

6.3. TRAJECTORY PRESCRIBED PATH CONTROL

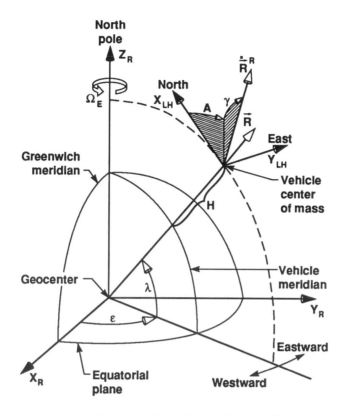

Figure 6.3.1: Differential Variables in Relative Coordinates

$$V'_R = \frac{-D}{m} - g\sin(\gamma)$$
$$- \Omega_E^2 r \cos(\lambda)\Big(\sin(\lambda)\cos(A)\cos(\gamma) - \cos(\lambda)\sin(\gamma)\Big)$$

$$\gamma' = \frac{L\cos(\beta)}{mV_R} + \frac{\cos(\gamma)}{V_R}\left(\frac{V_R^2}{r} - g\right) + 2\Omega_E \cos(\lambda)\sin(A) \quad (6.3.4)$$
$$+ \frac{\Omega_E^2 r \cos(\lambda)}{V_R}\Big(\sin(\lambda)\cos(A)\sin(\gamma) + \cos(\lambda)\cos(\gamma)\Big)$$

$$A' = \frac{L\sin(\beta)}{mV_R \cos(\gamma)} + \frac{V_R}{r}\cos(\gamma)\sin(A)\tan(\lambda)$$
$$- 2\Omega_E \Big(\cos(\lambda)\cos(A)\tan(\gamma) - \sin(\lambda)\Big)$$
$$+ \frac{\Omega_E^2 r \cos(\lambda)\sin(\lambda)\sin(A)}{V_R \cos(\gamma)}$$

where

$$\begin{aligned}
r &= H + a_e, \\
a_e &= \text{the earth radius}, \\
g &= \mu/r^2, \text{ the gravity force}, \\
\mu &= \text{gravitational constant}, \\
\Omega_E &= \text{earth angular rotation rate}, \\
m &= \text{the mass of the vehicle}, \\
L &= .5\rho C_L S V_R^2, \text{ the aerodynamic lift force}, \\
\rho(H) &= \text{the atmospheric density}, \\
C_L(\alpha) &= \text{the aerodynamic lift coefficient}, \\
S &= \text{the vehicle cross sectional reference area}, \\
D &= .5\rho C_D S V_R^2, \text{ the aerodynamic drag force}, \\
C_D(\alpha) &= \text{the aerodynamic drag coefficient}.
\end{aligned}$$

The control variables, which effectively determine the magnitude and direction of the aerodynamic force acting on the vehicle, are measured in a *body* coordinate system. The body coordinate system is defined in Figure 6.3.2. The bank angle β corresponds to a rotation or 'roll' about the vehicle's longitudinal axis (i.e., the X_B axis). The angle of attack α is measured from the relative velocity vector of the vehicle to the body X_B axis, and corresponds to a rotation or 'pitch' about the body Y_B axis. A more precise definition of these quantities and this coordinate system is given in [30].

The index of the resulting DAE system in relative coordinates can be found by a reduction technique based on repeated differentiations of the algebraic equations. In principle, this procedure reduces the index of the system to one. However, it may be difficult or impossible to apply in practice. There is also the previously mentioned tendency for the numerical solution of the reduced index problem to drift off the original path constraints.

Usually the differential equations are expressed in inertial coordinates, while the path constraints are written in terms of the relative differential variables. Since the local index is not invariant with respect to general transformations of variables, it is not clear whether the two DAE systems, one posed in inertial coordinates and the other in relative coordinates, have the same local index. The global index is unchanged by nonlinear coordinate changes [59].

Index Two TPPC Problem

We first consider a typical, but simplified, TPPC problem. Suppose it is desired to fly a vehicle along a prescribed azimuth and flight path angle

6.3. TRAJECTORY PRESCRIBED PATH CONTROL

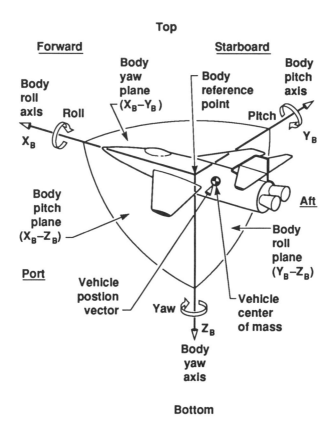

Figure 6.3.2: Body Coordinate System

trajectory given by the state variable constraints

$$\begin{aligned} \gamma + 1 + 9\left(\frac{t}{300}\right)^2 &= 0 \\ A - 45 - 90\left(\frac{t}{300}\right)^2 &= 0 \end{aligned}, \quad 0 \le t \le 300. \quad (6.3.5)$$

The flight path angle γ will range from $-1°$ to $-10°$, while the azimuth A will vary from $45°$ to $135°$. Both bank angle and angle of attack will be used as control variables. The DAE (6.3.4), (6.3.5) is index two for a lifting reentry vehicle as long as $(0.5\rho S V_R/m)^2(.01)C_L/\cos(\gamma) \ne 0$. On physical grounds we require $V_R \ne 0$, $\rho \ne 0$, $\gamma \ne \pi/2$, $\lambda \ne \pi$, and $\alpha \ne 0$. The related index one DAE can be derived directly by differentiating the algebraic equations (6.3.5) once and substituting for A' and γ' from the differential equations. Consistent initial values for the DAE system are determined by selecting the initial differential and control variables to satisfy (6.3.5) and the two new algebraic equations in the related index one system. The set of initial values used in our experiments were $H = 100,000$ ft., $\xi = 0°$, $\lambda = 0°$, $V_R = 12,000$ ft./sec., $\gamma = -1°$, $A = 45°$, $\beta = -.05220958616134°$, and $\alpha = 2.6728700742°$. These initial control variable values agree closely with values computed by Leimkuhler [161] using a symbolic mathematics package. He also computes initial derivative values for the controls, but in our numerical tests they are assumed to be zero. The initial derivative values for the differential variables are obtained from (6.3.4).

We now present the results of numerical experiments on this index two DAE. The computations were performed in single precision on a CDC 176 computer where the unit roundoff value is roughly 1.E-14. Numerical solutions to the index two problem, as well as the corresponding index one problem, were generated by the code DASSL discussed in Chapter 5. For this test problem, scalar relative (RTOL) and absolute (ATOL) error tolerances were specified, and ATOL was taken to be equal to RTOL. A truth model was constructed by applying DASSL with a tight local error tolerance (RTOL = 1.E-10) to the index one problem.

We have seen in Section 5.4.2 that the stepsize selection algorithm in DASSL is not designed to handle index two systems. Thus for the index two TPPC problem, the integration control tests were altered to exclude the algebraic variables. In spite of this modification to the error control, the accuracy of the algebraic variables turned out to be surprisingly good. Without this modification, the code fails repeatedly to accept a step until finally the minimum stepsize is reached and the algorithm aborts. An alternative approach to error control for the index two system would be to project the errors onto a different subspace, perhaps by a filtering technique as discussed in Section 5.4.2.

The index two and corresponding index one systems in relative coordinates were solved by DASSL. Plots of the control histories and of the prescribed paths are given in Figures 6.3.3 and 6.3.4, respectively. Relative errors in all variables at $t = 300$ are given in Table 6.3.1 for three values of the local error tolerance input parameter. Integration statistics on the number of integration steps (STEPS), evaluations of the differential equations (FNS) and Jacobian matrices (JACS), failed corrector iterations (CTF), and error test failures (ETF) are given in Table 6.3.2. For comparison purposes, integration statistics for the problem formulated as an index two system in inertial variables (6.3.3) are also given in Tables 6.3.1 and 6.3.2. For reference, the solution to the truth model at $t = 300$ yields the values $V_R = 1,433.29213$ ft./sec., $\gamma = -10°$, $H = 14,200.8114$ ft., $A = 135°$, $\lambda = 2.33149735°$, $\xi = 4.17108462°$, $\alpha = 7.15558457°$, $\beta = 26.3775452°$.

Performance Analysis of Space Vehicles

Typically, TPPC problems are formulated to aid in the performance analysis for the design of a space vehicle. A problem of major concern for lifting

6.3. TRAJECTORY PRESCRIBED PATH CONTROL

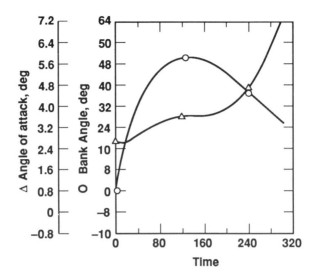

Figure 6.3.3: Control Histories

reentry vehicles is to determine their maximum crossrange (or downrange) capability. The trajectory profiles are constrained by skin temperature restrictions which result from the design of the vehicle's thermal protection system. A direct approach to determining a vehicle's maximum crossrange capability subject to heating constraints is given by optimal control theory, where the problem can be formulated as a two point boundary value problem involving DAE's. This approach introduces a set of adjoint variables for which there are associated differential equations. The DAE also includes the equations of motion and some algebraic equations arising from the optimality conditions. While a general method using a combination of a nonlinear programming technique and a shooting method has been proposed in [21] for solving such optimal control problems, its success is strongly dependent on having a starting solution which lies sufficiently close to the optimal solution. When heating constraints are included, the sensitivity of the trajectories to the initial guesses in the shooting problems is extreme, and the work in determining the optimal solutions can be exorbitant. Therefore, an indirect TPPC approach to determining a vehicle's crossrange capability may be more successful. Alternatively, a TPPC problem may be used to generate a 'good' initial solution for the optimal control formulation.

It is generally accepted that crossrange capability close to the maximum will result when the angle of attack is held at that value corresponding to the maximum lift/drag. NASA typically holds the angle of attack constant at about forty degrees in shuttle reentry simulations. This leaves bank angle to be used to shape the reentry trajectory to satisfy any remaining functional

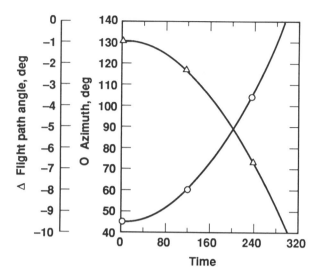

Figure 6.3.4: Prescribed Paths

constraints. In [32], a reentry vehicle was required to fly along a trajectory satisfying a drag-acceleration constraint. By varying certain parameters in the system, the crossrange capability of the vehicle could be evaluated, and effectively maximized. Posed in this way, the resulting TPPC system is a semi-explicit nonlinear *index three* DAE problem.

Index Three TPPC Problem

To illustrate some of the numerical problems encountered when solving higher index DAE's, such as initialization of variables, problem formulation, selection of a numerical method, and determination of the index, we consider this last example in more detail. We refer to the vehicle's position and velocity vectors as the vehicle's *state*. In the design of a safe reentry profile [32], mission restrictions concerning heating constraints and structural integrity requirements must be evaluated with respect to the forces acting on the vehicle. One way to accomplish this task is first to construct an acceptable drag acceleration corridor by analyzing the maximum load factors, the maximum surface temperature constraints at a set of control points, and the equilibrium glide boundaries. Then a *nominal* trajectory profile of the form

$$g(H, V_R, t) = 0 \qquad (6.3.6)$$

is selected inside this drag acceleration corridor. The first portion of the reentry phase entails flying the vehicle from a given state to one that lies on the nominal drag profile, and the second portion involves flying along this

6.3. TRAJECTORY PRESCRIBED PATH CONTROL

Variable	RTOL	Relative Coordinates		Inertial Coordinates
		Index 1	Index 2	Index 2
V_R	1.0E-6	8.76E-6	3.06E-6	2.73E-6
	0.5E-6	9.91E-7	5.30E-7	9.77E-8
	1.0E-8	4.19E-8	6.98E-9	0
γ	1.0E-6	4.20E-7	0	0
	0.5E-6	2.00E-7	0	0
	1.0E-8	1.00E-8	0	0
H	1.0E-6	3.20E-6	5.99E-7	2.18E-6
	0.5E-6	2.87E-6	2.79E-6	3.73E-7
	1.0E-8	9.15E-8	0	4.93E-8
A	1.0E-6	3.11E-7	0	0
	0.5E-6	1.48E-7	0	0
	1.0E-8	7.41E-9	0	0
λ	1.0E-6	1.37E-6	6.23E-6	7.33E-7
	0.5E-6	2.43E-6	4.89E-7	1.84E-7
	1.0E-8	5.58E-8	6.86E-8	3.00E-8
ξ	1.0E-6	3.96E-7	2.19E-6	3.98E-7
	0.5E-6	1.02E-6	2.16E-7	5.51E-8
	1.0E-8	1.68E-8	2.64E-8	1.68E-8
α	1.0E-6	1.43E-5	5.32E-6	4.52E-6
	0.5E-6	1.27E-7	2.64E-6	2.43E-5
	1.0E-8	2.66E-8	1.40E-8	7.97E-8
β	1.0E-6	8.02E-6	2.81E-6	9.78E-7
	0.5E-6	9.17E-7	4.85E-7	5.98E-5
	1.0E-8	4.17E-8	7.58E-9	1.48E-7

Table 6.3.1: Relative Errors for Prescribed Azimuth and Flight Path Angles

nominal drag profile. Figure 6.3.5 shows a typical drag acceleration profile as a function of the vehicle's relative velocity. The task now is to determine a control history which will cause the vehicle to fly along the drag acceleration profile. Choosing bank angle as the control variable, the required control history is shown in Figure 6.3.6.

We will consider only the first portion of the reentry trajectory. Equations (6.3.4) and (6.3.6) form an index three DAE system of the Hessenberg structure of size three (see Section 2.5.2), where $x_1 = (\gamma, A)$, $x_2 = (H, \lambda, \xi, V_R)$, and $x_3 = \beta$.

Let the particular path constraint have the form

$$\frac{D}{m} - \left(C_0 + C_1(V_R - V_0) + C_2(V_R - V_0)^2 + C_3(V_R - V_0)^3\right) = 0 \quad (6.3.7)$$

	INDEX	RTOL	STEPS	FNS	JACS	ETF	CTF
Truth Model	1	1.E-10	623	1552	52	25	0
Relative Coordinates	1	1.0E-6	155	414	22	3	0
		0.5E-6	152	404	24	3	0
		1.0E-8	324	810	33	12	0
Relative Coordinates	2	1.0E-6	66	233	24	2	1
		0.5E-6	73	253	28	0	3
		1.0E-8	113	400	27	3	1
Inertial Coordinates	2	1.0E-6	129	421	21	4	0
		0.5E-6	171	547	42	12	0
		1.0E-8	207	715	49	14	0

Table 6.3.2: Integration Statistics for Prescribed Azimuth and Flight Path Angles

where V_0 is the vehicle's initial velocity at t_0 and C_i ($i = 0, 1, 2, 3$) are constants. Since the vehicle's initial state is given at t_0 ($t_0 = 332.868734542$ sec.), the coefficients C_i ($i = 0, 1, 2, 3$) must be chosen so that this initial state is consistent for the DAE system. Since $V_R = V_0$ at $t = t_0$, we choose $C_0 = D/m$ and $C_1 = D'/(mV_0')$. Next, a point D^* on the nominal drag profile is selected to satisfy some additional physical requirements to prevent (6.3.7) from overshooting the nominal profile. The remaining two coefficients C_2 and C_3 are chosen so that (6.3.7) interpolates the point D^* and its slope on the nominal curve. If D^* is not chosen properly, the resulting cubic polynomial in (6.3.7) may overshoot the nominal drag profile before smoothly joining it. By choosing the coefficients in this way, the initial state vector is consistent with the algebraic equation (6.3.7) and its first derivative at t_0. To determine the initial bank angle, the second derivative of the algebraic equation must be satisfied. An ad-hoc numerical initialization scheme is used to obtain this value. Thus the initial value for bank angle contains a small error.

BDF codes such as DASSL or LSODI are written to vary the stepsize and order. However, from Section 3.2.4 we know that a constant stepsize must be used when solving an index three system of this type if the expected rate of convergence for the BDF methods is to be achieved. In particular, assuming consistent initial values are given, the $O(h)$ rate of convergence for the implicit Euler method is not attained in the algebraic variables until the second step. On the first step, $O(1)$ errors are expected in the algebraic variables. Since BDF codes such as DASSL and LSODI start the integration process with the first order method, their error control mechanisms detect the large errors. Thus these codes fail to complete the first integration step. Therefore, a simple constant stepsize implicit Euler code was written to generate the numerical solution to this index three system.

6.3. TRAJECTORY PRESCRIBED PATH CONTROL

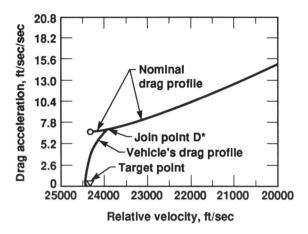

Figure 6.3.5: Drag Acceleration Profile

For consistent initial values, the index three system was solved by the implicit Euler method for a sequence of stepsizes. Since there is no analytic formula for the true solution, a truth model was computed by solving the corresponding index one DAE with DASSL using an extremely tight local error tolerance (RTOL = 1.E-12). Relative errors in the numerical solution of the index three system at $t = 419.868734542$ are given in Table 6.3.3. From Table 6.3.3 the global rate of convergence can be seen to be $O(h)$ in all variables.

Variable	Truth	h = .2	h = .1	h = .05
V_R	23816.7047	1.8E-5	8.9E-6	4.4E-6
γ	-.0968790848	1.9E-3	9.5E-4	4.7E-4
H	248938.336	9.5E-6	4.8E-6	2.4E-6
A	67.4966785	1.2E-5	6.2E-6	3.1E-6
λ	34.4379042	1.9E-5	9.3E-6	4.6E-6
ξ	183.872404	1.5E-6	7.6E-7	3.8E-7
β	61.5493127	1.6E-3	8.1E-4	4.1E-4

Table 6.3.3: Relative Errors in the Implicit Euler Solution
Index Three Drag Problem at $t = 419.868734542$

It is interesting to examine the local rates of convergence after only one step of the implicit Euler method. On the basis of the BDF theory for Hessenberg systems in Section 3.2.4, local rates of convergence are expected to be $O(h^2)$ in H, λ, ξ, and V_R (the x_2 variables), $O(h)$ in γ and A (the x_1 variables), and $O(1)$ in β [30]. The relative errors in the numerical solution and corresponding

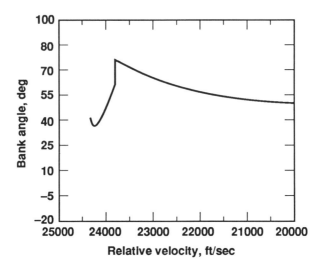

Figure 6.3.6: Bank Angle Control History

estimated rates of convergence for each variable are given in Table 6.3.4. Note that after the first step, the bank angle contains an $O(1)$ error as expected from the convergence theory. After the second integration step, β was $O(h)$ accurate.

Variable	h = .2	h = .1	h = .05	Order
V_R	3.8E-8	9.6E-9	2.4E-9	$O(h^2)$
γ	2.1E-3	1.0E-3	5.2E-4	$O(h)$
H	3.1E-7	7.8E-8	2.0E-8	$O(h^2)$
A	1.1E-5	5.5E-6	2.7E-6	$O(h)$
λ	3.7E-8	9.3E-9	2.3E-9	$O(h^2)$
ξ	5.3E-9	1.3E-9	3.3E-10	$O(h^2)$
β	1.2	1.2	1.2	$O(1)$

Table 6.3.4: Relative Errors in the Implicit Euler Solution
Index Three Drag Problem at $t = t_0 + h$

It is important to note that given the state vector, there are multiple solutions for the bank angle control. The k-step BDF method converges globally with kth-order accuracy after $k+1$ steps to a solution of the DAE provided the starting values are *numerically consistent to order* $k+1$ (see Section 3.2.4). It is possible for the BDF methods to produce a numerical solution which jumps from one solution curve to another during the first few steps, when the initial state values contain small errors. If the initial values are perturbed slightly in this example, then the implicit Euler solution for the control variable may

6.3. TRAJECTORY PRESCRIBED PATH CONTROL

converge to a control which is inconsistent with the given initial control value. This difficulty is the result of the nonuniqueness of the solution for the algebraic variable (due to the nonlinearity of the DAE), and the fact that the $O(1)$ error introduced into the algebraic variable during the first step may cause the numerical solution to jump to another solution curve.

Solution of this nonlinear index three system by a BDF method proved to be a challenging task, due to the sensitivity of the BDF to errors in the starting values. A code implementing the two-step BDF with a constant stepsize was applied to this index three system in [30]. If sufficiently accurate starting values were obtained, the two-step BDF determined the solution to $O(h^2)$ accuracy. However, generating these values using the implicit Euler method (as codes such as DASSL or LSODI would do) turned out to be a very sensitive procedure. Numerical experiments showed that slightly less accurate starting values could both destroy the global rates of convergence observed and cause the numerical solution for the control variable to jump to a different solution curve. These difficulties raise questions on the usefulness of the BDF methods for these types of index three systems.

Two singly-implicit two stage implicit Runge-Kutta (IRK) methods [38] were also applied [30] to this same control problem. The methods are SIRK22, a second order (for ODE's) L-stable method ($\lambda = (2 - \sqrt{2})/2$), and SIRK23, a third order A-stable method ($\lambda = (3 + \sqrt{3})/6$). Estimated rates of convergence for these two IRK methods are shown in Tables 6.3.5 and 6.3.6. From the tables, we can observe that both methods converge to $O(h)$ accuracy in the algebraic variable, while the differential variables remain at least second order accurate for both methods. These observations are in agreement with the recent convergence results of Hairer, Lubich and Roche [136] for index three Hessenberg systems (see Section 4.4).

Variable	h = 1.	h = .5	h = .25	Order
V_R	9.7E-10	1.3E-10	1.3E-11	$O(h^3)$
γ	2.0E-5	5.3E-6	1.4E-6	$O(h^2)$
H	2.2E-8	2.7E-9	3.4E-10	$O(h^3)$
A	4.0E-7	5.8E-8	9.3E-9	$O(h^{2+})$
λ	4.3E-8	5.7E-9	7.5E-10	$O(h^3)$
ξ	4.3E-9	5.4E-10	4.4E-11	$O(h^3)$
β	3.5E-4	1.7E-4	8.6E-5	$O(h)$

Table 6.3.5: Relative Errors in the SIRK23 Solution
Index Three Drag Problem at $t = 400.868734542$

The behavior of these IRK methods during the first few integration steps is interesting. Recall that the initial bank angle is not given exactly. For this nonlinear index three system, the A-stable method SIRK23 with stability

Variable	h = 1.	h = .5	h = .25	Order
V_R	5.6E-8	1.4E-8	3.5E-9	$O(h^2)$
γ	9.6E-5	2.4E-5	6.0E-6	$O(h^2)$
H	6.7E-8	1.7E-8	4.2E-9	$O(h^2)$
A	1.2E-6	3.1E-7	7.5E-8	$O(h^2)$
λ	7.3E-8	1.9E-8	4.4E-9	$O(h^2)$
ξ	1.1E-8	2.7E-9	6.2E-10	$O(h^2)$
β	1.4E-2	7.0E-3	3.5E-3	$O(h)$

Table 6.3.6: Relative Errors in the SIRK22 Solution
Index Three Drag Problem at $t = 400.868734542$

constant $r \approx -.73$ produces β values which oscillate about the correct value, with these oscillations gradually being damped out. On the other hand, the L-stable ($r = 0$) method SIRK22 annihilates the initial error in β on the first integration step. An advantage in applying IRK methods to higher index systems is that they do not require additional starting values.

From this example, we see that IRK methods, and in particular the L-stable formulas, may be useful not only in determining starting values for the BDF, but also as an class of alternative numerical methods for solving index three systems. As the application of BDF methods to higher index DAE's is often not feasible in practice due to their sensitivity to errors in the starting values and to changes in stepsize and order, IRK methods represent a viable alternative.

6.4 Electrical Networks

Historically, computer aided design of electrical networks stimulated researchers to study DAE's and their direct numerical solution. In fact, the first paper on the numerical solution of DAE's appeared in the IEEE Transactions on Circuit Theory in 1971 [113]. In this paper, C. W. Gear introduced BDF methods for the solution of differential-algebraic systems, and for many years they were the only numerical methods considered in the literature. The application of the BDF methods to large scale *descriptor* systems, the more common name for DAE's in circuit analysis, was proposed again in 1978 by Manke et al.[179] for several reasons. First, the BDF methods are able to exploit the natural sparsity of the system, which is generally destroyed if techniques to reduce the DAE to a *state variable system* (an explicit ODE system) are employed. Secondly, for large descriptor systems the reduction techniques, including the shuffle algorithm, the Q-Z algorithm, and methods based on singular value decompositions, are no longer practical. In fact, it is not always possible to derive a state variable representation [71]. Therefore research began on the de-

6.4. ELECTRICAL NETWORKS

velopment of numerical methods that could be applied directly to large scale descriptor systems. DAE's arising in this field exhibit a natural structure somewhat different from what we have seen earlier in DAE's arising from either mechanical systems or in trajectory control problems. In particular, these DAE's tend to be large and sparse. Frequently they are also linear, although some circuit devices introduce nonlinearities. Our interest here is not to discuss the design of circuits in detail, nor to give numerical results, but rather to show what kind of DAE's may arise and what aspects of the circuit affect such DAE properties as index and solvability.

Consider an electrical network composed of B branches connected to N nodes. Assigning a current variable i_B to each branch and a voltage variable v_N to each node, the circuit equations are derived from Kirchoff's laws:

1. The algebraic sum of the currents into a node at any instant is zero.

2. The algebraic sum of the voltage drops around a loop at any instant is zero.

By convention, current is the net flow of positive charge, so we designate a current direction along each branch by assigning one node to be negative and the other node to be positive (the current flow being from positive to negative). The topology of the circuit can now be described via a $B \times N$ *network incidence matrix* A. The (i,j) element of A is ± 1 if node j is the $\pm$ node for the ith branch. Letting i_B denote the vector of current variables, Kirchoff's current law simply states that $A^T i_B = 0$. The voltage drop across each branch is defined to be the difference between the voltage at the positive node and the voltage at the negative node. These branch voltages v_B can be expressed in terms of the nodal voltages v_N as $v_B = Av_N$.

First, we consider circuits with branches containing capacitors, inductors and resistors. Energy is stored in the form of a charge or an electrical field in a capacitor, while in an inductor it is stored in the form of a magnetic field. Resistors are used to effect a loss or gain of power in a branch. For now, we assume the relationship between current and voltage across branches with these elements is linear. More precisely, the voltage-current or *v-i* relationship (*characteristic*) across a branch with a resistor is assumed to satisfy Ohm's law, $v_R = Ri_R$ with positive resistance R. The *v-i* characteristics of a linear capacitor and a linear inductor satisfy $i_C = Cv'_C$ and $v_L = Li'_L$, respectively. Networks with linear resistors, capacitors, and inductors are commonly referred to as linear RLC circuits. There may also be voltage sources where $v_E = e(t)$ for any current i_E, and current sources where $i_S = i(t)$ for any voltage v_S. We partition the incidence matrix, the branch currents, and the branch voltages as follows

$$A = \begin{bmatrix} A_E \\ A_C \\ A_R \\ A_L \\ A_S \end{bmatrix}, \quad i_B = \begin{bmatrix} i_E \\ i_C \\ i_R \\ i_L \\ i_S \end{bmatrix}, \quad v_B = \begin{bmatrix} v_E \\ v_C \\ v_R \\ v_L \\ v_S \end{bmatrix}. \qquad (6.4.1)$$

The subscripts R, C, L, E, and S denote the types of element (resistor, capacitor, inductor, voltage source or current source respectively), that are present on the branch. The circuit equations can then be written as

$$Hx' + Gx = f \qquad (6.4.2)$$

where

$$H = \text{diag}\,[0, C, 0, 0, 0, 0, 0, 0, L, 0, 0],$$

$$G = \begin{bmatrix} I & 0 & 0 & 0 & 0 & 0 & 0 & 0 & 0 & 0 & 0 \\ 0 & 0 & 0 & 0 & 0 & 0 & -I & 0 & 0 & 0 & 0 \\ 0 & 0 & I & 0 & 0 & 0 & 0 & -R & 0 & 0 & 0 \\ -I & 0 & 0 & 0 & 0 & 0 & 0 & 0 & 0 & 0 & A_E \\ 0 & -I & 0 & 0 & 0 & 0 & 0 & 0 & 0 & 0 & A_C \\ 0 & 0 & -I & 0 & 0 & 0 & 0 & 0 & 0 & 0 & A_R \\ 0 & 0 & 0 & -I & 0 & 0 & 0 & 0 & 0 & 0 & A_L \\ 0 & 0 & 0 & 0 & -I & 0 & 0 & 0 & 0 & 0 & A_S \\ 0 & 0 & 0 & -I & 0 & 0 & 0 & 0 & 0 & 0 & 0 \\ 0 & 0 & 0 & 0 & 0 & 0 & 0 & 0 & 0 & I & 0 \\ 0 & 0 & 0 & 0 & 0 & A_E^T & A_C^T & A_R^T & A_L^T & A_S^T & 0 \end{bmatrix},$$

$$x = \begin{bmatrix} v_B \\ i_B \end{bmatrix}, \quad f = \begin{bmatrix} e(t)^T, 0, 0, 0, 0, 0, 0, 0, 0, i(t)^T, 0 \end{bmatrix}^T.$$

The inductance matrix L is diagonal if the inductors are uncoupled, and positive semidefinite if they are mutually coupled. R and C are diagonal matrices with positive diagonal elements, and I is the identity matrix. Therefore system (6.4.2) is generally quite sparse and transformable to semi-explicit form by a reordering of the variables. The formulation (6.4.2) is still valid if H and G depend on t.

Now consider a linear circuit composed only of resistors, capacitors and voltage sources. For simplicity, let the resistance and capacitance matrices, R and C, be constant. Suppose that there is a path from every node along branches to the ground node, that is the circuit is *connected*. This assumption is equivalent to assuming that the columns of the incidence matrix A are linearly independent so that the rank of A is N and $N \geq B$. Note that two rows of A are linearly dependent if and only if they are identical since each row contains only two nonzero entries, $+1$ and -1. In this case, the circuit contains a two branch loop between these two nodes, and one branch could

6.4. ELECTRICAL NETWORKS

effectively be eliminated from the circuit. For a linear RC circuit with voltage sources, the circuit equations form an index one DAE providing the columns of the matrix A_* and the rows of A^* are linearly independent [179], where

$$A_* = \begin{bmatrix} A_E \\ A_C \\ A_R \end{bmatrix}, \quad A^* = \begin{bmatrix} A_E \\ A_C \end{bmatrix}.$$

In other words, the circuit must be connected, the voltage sources independent, and there must be no loops containing only capacitors. The reduction to state variable form can be accomplished by straightforward algebraic computation, hence the index is one. However, the sparsity of the original DAE is generally destroyed by this process. In general, the shuffle algorithm [174,175,179] can be used to transform a linear constant coefficient DAE to state variable form whenever its matrix pencil is regular (that is, the DAE is solvable in the sense of Section 2.3). Note that the solvability of these simple circuits depends only on the topology of the network, and not on the specific resistance and capacitance values so that it is a structural property.

If A^* has linearly dependent rows in a linear RLC circuit, then the circuit equations form an index two DAE system [170]. The reduction of the DAE to state variable form cannot be accomplished by purely algebraic computations, but will in fact require one differentiation of a subset of algebraic equations corresponding to the voltage equations relating the nodal and branch voltages. In this case, the corresponding state variable system involves the first derivative of the voltage source $e(t)$. Two iterations of the shuffle algorithm are now required [179].

While linear models often suffice, many electrical devices are not linear. For example, resistors may satisfy a nonlinear v-i relationship such as $v_R = f(i_R, t)$, or other components such as operational amplifiers or transistors may be included in the circuit design.

Arbitrarily high index DAE's can arise in circuits containing differential amplifiers, which can be realized using operational amplifiers. An operational amplifier is a three terminal device which has two input terminals and one output terminal, as shown in Figure 6.4.1. In an ideal operational amplifier, there is no voltage drop or current across the input branch, and the gain in the amplifier is said to be infinite. For example, the operational amplifier in Figure 6.4.1 satisfies the relation $v_3 = A(v_1 - v_2)$ for $A \approx 10^5$. Now consider a circuit with one differential amplifier [50,179] as shown in Figure 6.4.2. Assuming an ideal operational amplifier, the circuit equations lead to the relation $v_3 = -CRe'$ where e is the voltage source. Since the solution to the circuit equations involves at least one derivative of the input function, the index of this system must be at least two. By cascading a series of differential amplifiers in a circuit as in Figure 6.4.3, the index of the resulting DAE can be made arbitrarily high. The index of the DAE corresponding to Figure 6.4.3 is at least four since $v_3 = -C_3 R_3 C_2 R_2 C_1 R_1 e'''$.

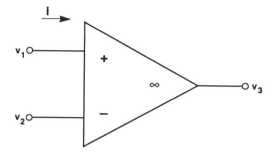

Figure 6.4.1: Operational Amplifier

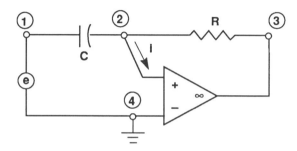

Figure 6.4.2: Circuit with One Differential Amplifier

The solvability of DAE's arising from circuits with operational amplifiers depends not only on the topology of the network, but also on the specific element values. As an example, consider the circuit composed of four resistors, an operational amplifier and a voltage source [179] in Figure 6.4.4.

The circuit equations are

$$\begin{bmatrix} G_1 & -G_1 & 0 & 0 & 1 & 0 \\ -G_1 & G_1+G_2 & -G_2 & 0 & 0 & 0 \\ 0 & -G_2 & G_2+G_4 & -G_4 & 0 & 1 \\ 0 & 0 & -G_4 & G_3+G_4 & 0 & 0 \\ 1 & 0 & 0 & 0 & 0 & 0 \\ 0 & 1 & 0 & -1 & 0 & 0 \end{bmatrix} \begin{bmatrix} v_1 \\ v_2 \\ v_3 \\ v_4 \\ i_E \\ i_o \end{bmatrix} = \begin{bmatrix} 0 \\ 0 \\ 0 \\ 0 \\ e \\ 0 \end{bmatrix} \quad (6.4.3)$$

where $G_i = 1/R_i, i = 1, 2, 3, 4$ and i_0 denotes the output current of the operational amplifier. Clearly the coefficient matrix must be nonsingular in order for this purely algebraic linear system to be solvable. In other words, its determinant,

$$\frac{1}{R_2 R_3} - \frac{1}{R_1 R_4}$$

must be nonzero. Hence, solvability of this system depends on the specific values of the resistances.

6.4. ELECTRICAL NETWORKS

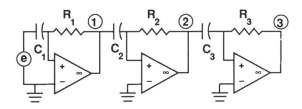

Figure 6.4.3: Circuit with Three Differential Amplifiers

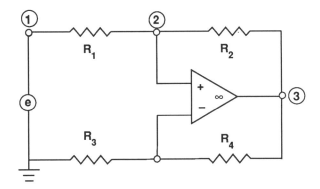

Figure 6.4.4: Solvability of Circuit with Operational Amplifier

The number of state variables in a circuit with operational amplifiers may also depend on the particular values given to the circuit components. For example, replace the voltage source in the circuit given in Figure 6.4.4 with a capacitor. The circuit equations are now

$$\begin{bmatrix} Cv_1' \\ 0 \\ 0 \\ 0 \\ 0 \end{bmatrix} + \begin{bmatrix} G_1 & -G_1 & 0 & 0 & 0 \\ -G_1 & G_1+G_2 & -G_2 & 0 & 0 \\ 0 & -G_2 & G_2+G_4 & -G_4 & 1 \\ 0 & 0 & -G_4 & G_3+G_4 & 0 \\ 0 & 1 & 0 & -1 & 0 \end{bmatrix} \begin{bmatrix} v_1 \\ v_2 \\ v_3 \\ v_4 \\ i_o \end{bmatrix} = \begin{bmatrix} 0 \\ 0 \\ 0 \\ 0 \\ 0 \end{bmatrix} \quad (6.4.4)$$

The determinant of the matrix pencil of system (6.4.4) is

$$\lambda C(G_1G_4 - G_2G_3) - G_1G_2G_3,$$

which for nonzero resistances is never identically zero. Now the number of state variables (in other words, the *core-rank* of the DAE [49]) is equal to the degree of the polynomial $\det(\lambda G+H)$. Equivalently, this is the number of independent initial conditions which must be specified. Thus, if $G_1G_4 - G_2G_3 = 0$, the only solution is the zero solution. If $G_1G_4 - G_2G_3 \neq 0$, there is a one parameter family of solutions to (6.4.4). Therefore, the number of state variables are determined not only by the topology of the circuit, but also by the specific resistance values.

DAE's in the form
$$Ax' + B(x) = f(t)$$
or
$$A(x)x' + B(x) = f(t) \tag{6.4.5}$$
where A and B are nonlinear arise from circuits which include transistors or unicursal elements [50,72], respectively. A unicursal curve is one that can be continuously parameterized by a function defined on a connected interval of the real line and which does not retrace any part of the curve. The parameterization is allowed to intersect itself. An element is unicursal if its v-i characteristic is a unicursal curve. Transistors are often modeled by linear capacitors and nonlinear resistors. Resistors, inductors, or capacitors may be characterized by unicursal curves. An example of a single loop circuit with no sources leading to a DAE in the form (6.4.5) is given in [72]. For the parameters τ_1, τ_2, and τ_3, the v-i characteristic of resistance is defined by

$$v_R = \cos^3(\tau_1), \quad i_R = \sin^3(\tau_1),$$

its flux-current characteristic of the inductance is given as

$$\phi_L = \cos(\tau_2), \quad i_L = \sin(\tau_2),$$

and the charge-voltage characteristic of the capacitance is

$$q_C = \tau_3 - \sin(\tau_3), \quad v_C = 1 - \cos(\tau_3).$$

The circuit equations can be written in terms of the parameters as

$$\begin{bmatrix} \sin(\tau_2) & 0 & 0 \\ 0 & 1 - \cos(\tau_3) & 0 \\ 0 & 0 & 0 \end{bmatrix} \begin{bmatrix} \tau_2' \\ \tau_3' \\ \tau_1' \end{bmatrix} = \begin{bmatrix} \cos^3(\tau_1) - \cos(\tau_3) + 1 \\ 2\sin(\tau_2) \\ \sin^3(\tau_1) - 2\sin(\tau_2) \end{bmatrix}.$$

In summary, linear RLC circuits can lead to large sparse linear DAE's, while circuits with transistors or unicursal elements tend to make the DAE systems nonlinear. Solvability of DAE's arising from linear circuits without operational amplifiers is determined by the network topology alone. However, linear circuits with operational amplifiers lead to DAE's whose solvability depends on the particular v-i characteristics of the circuit components. Similarly, the number of independent initial conditions may also depend on the specific values of resistances, capacitances, etc. That the index of DAE's arising from circuits may be arbitrarily high is clear from the example of a cascade of differential amplifiers. Higher index DAE's can also occur if different variables are chosen as inputs and outputs. For example, whether a device is a differentiator or an integrator depends on what is considered an output and what is considered an input.

DAE's have also been used to model hysteresis in circuits [192]. The DAE formulation has the advantage that one DAE models all of the circuit behavior, whereas different ODE models are needed for different parts of the hysteresis loop.

6.5 DAE's Arising from the Method of Lines

In Chapter 1 we pointed out how DAE's arise when the method of lines (MOL) is used to solve partial differential equations. Here we will study this approach further, to illustrate the advantages of a DAE formulation, how consistent initial conditions are determined, and how problem formulation plays an important role in the success or failure of the numerical solution process.

The method of lines is applied most often to parabolic systems of PDE's. The underlying idea of this approach is to discretize the spatial derivatives by finite differences or finite elements, to obtain an initial value ODE system, or more generally a DAE system. If the resulting system is an explicit ODE, $y' = f(t, y)$, robust and computationally efficient software is widely available which can be applied to generate a numerical solution. Many MOL problems lead naturally to either implicit ODE's or DAE's in the form $F(t, y, y') = 0$. While these systems can sometimes be manipulated into an explicit ODE form, a more direct approach is to apply software designed for DAE's. As software for DAE's has been widely available only recently [143,200], there have been fewer practitioners of the DAE approach. However, examples of this more general approach can already be found in applications arising in combustion and chemical kinetics modeling [148,152], adaptive moving grid methods [203], convective flow in the earth's mantle [207], gas transfer in piping networks [243,244], and chemical vapor deposition for the fabrication of integrated circuits [28,29,80,81]. Here we investigate further the examples that were first discussed in Chapter 1.

Combustion Ignition

We first consider an example from combustion ignition [176,208]. The specific problem is one in which a quiescent uniform gas mixture is contained in a closed cylindrical vessel, as in Figure 6.5.1. At the start of the problem an external source (such as a laser) deposits energy near the vessel's center for a prescribed period of time, after which the source is no longer applied. Depending on the gas composition and the strength of this external energy source, a flame may propagate from the source and burn all the gas. The objective of such a simulation is to gain a better understanding of the ignition process, for example by determining the minimum external energy source required to cause ignition.

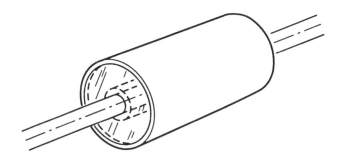

Figure 6.5.1: Ignition of Combustible Gases

For our purpose, it is sufficient to study the ignition of a single-component nonreacting gas in a closed cylindrical vessel, as in Kee and Petzold [151]. It is assumed the ignition source is weak, the sound speed is high, and the pressure disturbances are time varying, but spatially uniform. Using time t and a Lagrangian coordinate ψ as the independent variables, the equations modeling this process are given by [151]

$$\frac{\partial T}{\partial t} - \frac{1}{\rho c_p} \frac{\partial p}{\partial t} = \frac{1}{c_p} \frac{\partial}{\partial \psi} \left(\rho r^2 \lambda \frac{\partial T}{\partial \psi} \right) \qquad (6.5.1a)$$

$$0 = \frac{\partial r}{\partial \psi} - \frac{1}{\rho r} \qquad (6.5.1b)$$

$$0 = \frac{\partial p}{\partial \psi} \qquad (6.5.1c)$$

$$0 = p - \rho \frac{RT}{\overline{W}}. \qquad (6.5.1d)$$

Equation (6.5.1a) represents conservation of energy, (6.5.1b) defines the physical coordinate r, (6.5.1c) is the uniform pressure condition, and (6.5.1d) is the equation of state. Dependent variables in these equations are the temperature T, the pressure p, the density ρ, and the spatial coordinate r. Constants, dependent only on the gas mixture, include the mean molecular weight $\overline{W}$, the thermal conductivity λ, and the constant pressure heat capacity c_p. R is the universal gas constant. The total mass in the vessel can be defined as

$$\psi_R = \int_0^R \rho r \, dr,$$

where ρ is the initial density and R is the radius of the vessel. Assuming that no mass can exit or leave the vessel, ψ ranges from zero at the vessel center ($r = 0$) to ψ_R at the outer wall ($r = R$). Since (6.5.1d) can be used to

6.5. DAE'S ARISING FROM THE METHOD OF LINES

evaluate the density ρ wherever it occurs, we consider the DAE obtained by spatial discretization (i.e., in the Lagrangian variable ψ) of (6.5.1a)-(6.5.1c) with the three dependent variables r, T, p. Thus, there are three equations and three unknowns. The system has order four since the energy equation is second order, while the uniform pressure condition and the physical coordinate equation are first order. Therefore, four boundary conditions for the dependent variables are required. In addition, consistent initial values must be specified.

Note that the energy equation has two time derivatives, while the other two equations have none. Therefore, after discretizing the spatial derivatives in the system, the new equations arising from (6.5.1a) are implicit in the time derivatives, while those arising from (6.5.1b) and (6.5.1c) are purely algebraic. The solution of this new system with standard ODE software would require first that the system be reformulated into the required form $y' = f(t, y)$. However, this system is already a DAE.

Initial conditions need to be selected for this DAE. Typically, the initial profile for temperature and pressure are specified. An initial value for r will also be required since it is now a dependent variable. Given p and T, r can then be chosen uniquely to satisfy the integral of equation (6.5.1b), namely

$$r = \int_0^{\psi_R} \frac{1}{\rho r} d\psi. \tag{6.5.2}$$

In implementing (6.5.2) to find the initial values for r, it is important to select a quadrature scheme for (6.5.2) which is consistent with the spatial differencing of equation (6.5.1b). Otherwise, the numerical DAE method, which requires consistent starting values, is likely to fail.

Next, the boundary conditions must be specified. In physical terms, the vessel walls are held at a constant temperature T_w, and mass must be conserved. In terms of the Lagrangian independent variable, these boundary conditions can be interpreted as requiring

$$r = 0, \quad \frac{\partial T}{\partial \psi} = 0$$

at the center of the vessel ($\psi = 0$), and

$$r = R, \quad T = T_w$$

at the vessel boundary ($\psi = \psi_R$). Note that we have four boundary conditions, two on r and two on T, but none on p.

The spatial discretization of equations (6.5.1a)-(6.5.1c) may be done in many different ways. Kee and Petzold used forward differencing for the uniform pressure equation (6.5.1c), backward differencing for the physical coordinate equation (6.5.1b), and second order central differences for the second order derivatives in the energy equation (6.5.1a). For example, (6.5.1c) is discretized for a fixed spatial grid in ψ as

$$\frac{p_{j+1} - p_j}{\psi_{j+1} - \psi_j} = 0, \quad j = 0, 1, \ldots, N - 1.$$

Note that this scheme effectively requires that $p_0 = p_1 = \ldots = p_N$, so that in some sense there is an implicit boundary condition imposed on p. (Recall that it is assumed that p depends only on t.)

The numerical scheme just described leads to a nonlinear index one DAE which is linear in the time derivatives. The selection of consistent initial values for the dependent variables is done in conjunction with the choice of spatial discretization, in order to insure numerical consistency of the DAE equations. Boundary conditions are simply treated as additional algebraic equations which the solution must satisfy.

It is useful to note that different formulations of DAE systems can lead to different Jacobian matrix structures. This is particularly important when implementing method of lines solutions to PDE's using a code like DASSL, which has an option for banded linear systems. Banded systems are advantageous over some other structures on vector computers, and optimized code for solving these linear systems is often available on supercomputers. In the case of the combustion ignition problem that we have been discussing, note that (6.5.1b) and (6.5.1c) were solved in their stated form, rather than treating the pressure as a global variable and evaluating r by an integral. The variable r could have been considered to be a coefficient in the diffusion term of the energy equation, and then it could have been evaluated from the integral equation (6.5.2). However, doing so expands the bandwidth of the Jacobian and results in an inefficient solution procedure. Therefore r is introduced as a dependent variable at each mesh point, and it is solved for with all the other variables. The system of equations is larger, but in a form more amenable to efficient solution. A similar argument is made for the evaluation of pressure p. The pressure appears in every other equation because it is needed to evaluate the density ρ from the equation of state. By solving (6.5.1c), pressure is introduced as a variable at each mesh point, and a trivial equation is solved to insure that all pressures are equal. Once again this allows the Jacobian matrix to be banded, but at the expense of creating more dependent variables. Using these ideas, it is often possible to avoid writing special purpose linear algebra software and instead use the standard software such as LINPACK [100].

The advantages of a DAE formulation for a similar, but more complicated, MOL problem arising from the boundary layer equations modeling the steady flow of a compressible liquid in a pipe are discussed in [152]. Initial profiles of velocity, temperature, and pressure may be selected arbitrarily, but the radial velocity must be chosen to satisfy a complicated equation obtained by substituting all the other equations into the continuity equation. This initialization problem can be seen more easily when the equations are rewritten to utilize the von Mises coordinate transformations. Then there is only one profile for the radial coordinate r, now a dependent variable in the system, which is consistent with the other initial conditions. As in our earlier example, the application of standard ODE software would be difficult here because two boundary conditions are given for r, but none are given for the pressure

6.5. DAE'S ARISING FROM THE METHOD OF LINES

p. In general, the flexibility of the DAE formulation allows the boundary conditions to be complicated nonlinear functions of all the dependent variables. The momentum equation involves two time-like derivatives, so that the DAE system which results from the MOL discretization is fully-implicit.

Incompressible Fluid Flow

As our next example, we investigate further the DAE system from Chapter 1 which arose when the MOL was applied to the equations modeling the flow of an incompressible viscous fluid (1.3.14a) and (1.3.14b) (the Navier-Stokes equations) [128]

$$MU' + (K + N(U))U + CP = f(U, P) \quad (6.5.3a)$$
$$C^T U = 0. \quad (6.5.3b)$$

Recall that $U(t)$ and $P(t)$ approximate the velocity $u(x,t)$ and pressure $p(x,t)$, respectively, on a fixed spatial mesh (in two or three dimensions). If finite differences are used to discretize the spatial derivatives, then the mass matrix M is an identity matrix, while for finite elements, it is a symmetric positive definite matrix. The DAE system (6.5.3) has local and global index two [170] if either

$$\frac{\partial f}{\partial P} = 0$$

and C has linearly independent columns, or if

$$C^T M^{-1} \left(C - \frac{\partial f}{\partial P} \right)$$

is nonsingular.

For a variety of reasons, which we have previously discussed in Chapter 5, the numerical solution of index two DAE's with currently available software for DAE's is not straightforward. In practice, index reduction methods have been employed whenever possible to reformulate these DAE's as index one problems. Standard DAE software such as DASSL [200] or LSODI [143] can then be applied in a straightforward way. In this example, one iteration of the reduction technique described in [123] can be applied to obtain the alternate index one DAE [129]

$$MU' + (K + N(U))U + CP = f(U, P) \quad (6.5.4a)$$
$$AP = C^T M^{-1} (f - (K + N(U))U), \quad (6.5.4b)$$

where $A = C^T M^{-1} C$. The reduction scheme involves one differentiation of the algebraic equation (6.5.3b) with respect to time to obtain $C^T U' = 0$, a premultiplication of (6.5.3a) by $C^T M^{-1}$, and a simplification to obtain (6.5.4b). Then (6.5.4a) can be solved with an explicit method, while (6.5.4b) can be solved by an implicit method. One disadvantage to this approach is that the

DAE system may no longer be sparse if finite elements were used in the MOL, although it is sparse for finite differences. Generally, one must be concerned about the 'drift' in the numerical solution of the reduced index DAE. However in this case the original algebraic constraint (6.5.3b) will still be satisfied by the numerical solution of the index one problem for many numerical methods since the constraint (6.5.3b) is linear with constant coefficients.

Adaptive Moving Meshes

The use of a fixed spatial mesh is not always practical, particularly in problems with rapidly moving fronts, such as in combustion modeling. In this case, the solution at a fixed mesh point may change very slowly except for a brief period when the front passes. To follow the front accurately, a very fine fixed mesh is required, which may be computationally prohibitive. Recently, several authors have proposed schemes that adjust the spatial mesh in order to resolve the front accurately as it moves. We study here a class of methods [141, 146,150] which move the mesh so that in transformed coordinates the solution varies less rapidly in time. This strategy involves inserting and deleting mesh points in order to resolve the spatial gradients. The original schemes proposed required the PDE's to have the time derivatives given explicitly as either

$$u_t + \nabla \cdot G(u) = S$$

in [150], or as

$$u_t = f(u, u_x, u_{xx})$$

in [146]. Petzold [203,204] has generalized the technique of moving the mesh to apply to the more general form

$$F(u_t, u, u_x, u_{xx}) = 0. \tag{6.5.5}$$

Suppose we select the grid velocity x' to minimize the time rate of change of u and x in the new moving coordinate system, that is

$$\min_{x'}(\| u' \|^2 + \alpha \| x' \|^2),$$

where α is a positive scalar weight. Noting that

$$\frac{du(x,t)}{dt} = \frac{\partial u(x,t)}{\partial x}\frac{dx}{dt} + \frac{\partial u(x,t)}{\partial t},$$

the velocity $u(x,t)$ and the mesh $x(t)$ then satisfy the PDE's

$$F(u' - u_x x', u, u_x, u_{xx}) = 0 \tag{6.5.6a}$$
$$\alpha x' + u' u_x = 0. \tag{6.5.6b}$$

The discretization of the spatial derivatives in (6.5.6) leads to a DAE system

6.5. DAE'S ARISING FROM THE METHOD OF LINES

where the time derivatives appear implicitly. Note that even if (6.5.5) were explicit as in $u_t = f(u, u_x, u_{xx})$, then the DAE (6.5.6) would be linear in the derivatives, but still not semi-explicit.

This moving mesh system illustrates the importance of problem formulation. Petzold [203,204] considered three different analytically equivalent formulations for the model problem $u_t = f(u, u_x, u_{xx})$ with $\alpha = 1$.

Case 1:
$$\begin{aligned} u' - u_x x' &= f(u, u_x, u_{xx}) \\ x' + u_x u' &= 0. \end{aligned} \quad (6.5.7)$$

Case 2:
$$\begin{aligned} u' - u_x x' &= f(u, u_x, u_x x) \\ x' &= -f(u, u_x, u_{xx}) u_x / (1 + u_x^2). \end{aligned} \quad (6.5.8)$$

Case 3:
$$\begin{aligned} u' &= f(u, u_x, u_{xx}) - u_x(f(u, u_x, u_{xx}) u_x / (1 + u_x^2)) \\ x' &= -f(u, u_x, u_x x) u_x / (1 + u_x^2). \end{aligned} \quad (6.5.9)$$

Note that only algebraic manipulations (no differentiations) of the equations in Case 1 are required to obtain the equations in either Case 2 or 3. Since Case 3 is simply an explicit ODE system, it follows that the equations in Cases 1 and 2 are implicit ODE's (index zero DAE's). While all three forms are analytically equivalent, the conditioning of the iteration matrices involved in the linear systems may be quite different. In particular, for problems with very steep fronts (u_x is large for some nodes), the condition number of the iteration matrix for (6.5.8) is much larger than the condition number corresponding to either (6.5.7) or (6.5.9).

To demonstrate the significance of these different formulations, we summarize the numerical results found in [203], where DASSL is used to solve

$$\begin{aligned} T_t &= T_{xx} + D(1 + \alpha - t)e^{-\delta/T}, \quad t > 0, \ 0 < x < 1 \\ T_x(0, t) &= 0, \quad T(1, t) = 1 \\ T(x, 0) &= 1, \quad 0 \leq x \leq 1 \end{aligned}$$

with $D = Re^\delta/(\alpha\delta)$. This problem is a simple model of a single step reaction with diffusion studied by Adjerid and Flaherty [1]. The temperature, initially one for all x in the domain, gradually increases with a maximum occurring at $x = 0$. Ignition occurs at a finite time, and the temperature at $x = 0$ increases rapidly to $1 + \alpha$. A steep front then forms and propagates toward $x = 1$ with a speed proportional to $.5e^{\alpha\delta}(1 + \alpha)^{-1}$. Once the flame reaches $x = 1$, the solution reaches a steady state. The numerical results obtained were obtained for $\alpha = 1$, $\delta = 30$, and $R = 5$ on the time interval $[0.0, 0.29]$. The moving mesh used a maximum of approximately thirty nodes. The solution curves, at time intervals of .0029, are shown in Figure 6.5.2. In Figure 6.5.3 the results of each time step for the appropriate spatial mesh are given. There it is clear how the mesh is moving, and how points are being added and deleted over time.

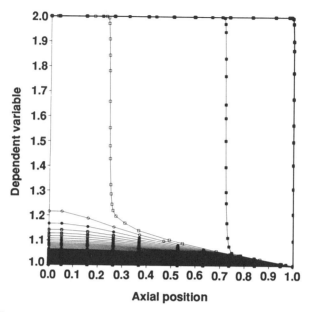

Figure 6.5.2: Single Step Reaction - Fixed Time Intervals

DASSL was used to integrate the three different formulations (Cases 1-3) for this test problem. All computations were performed in single precision on a Cray-1. The solutions found for each case are all of comparable accuracy. The differences due to the problem formulation show up primarily in the integration statistics which are given in Table 6.5.1. Here NODES is the maximum number of grid points used by the code over all time steps, STEPS is the number of time steps needed to complete the problem, FNS is the number of evaluations of the DAE that DASSL used, and JACS is the number of Jacobian evaluations. ETF is the number of times the time step was reduced due to a failure of the error test in DASSL, and CTF is the number of times the time step was reduced due to a failure of the Newton iteration in DASSL to converge. The results in Table 6.5.1 clearly demonstrate that the second problem formulation is a poor choice. In this case, ill-conditioning of the iteration matrix is responsible for the difficulties experienced by DASSL. For comparison, we also give the results when a fixed spatial mesh is used (Case 4). Note that the solutions for any of the moving mesh cases (Cases 1-3) were obtained more efficiently than the solution determined for a fixed spatial mesh.

It is not our intention to discuss here all the important concerns in computing adaptive moving mesh solutions. For example, the addition of a penalty term to the system, in order to prevent adjacent mesh points from colliding, is shown in [202] to reduce the number of time steps and function evaluations significantly. Also, the question of whether solvers other than DASSL may be able to integrate this problem more efficiently needs to be addressed. Finally,

6.5. DAE'S ARISING FROM THE METHOD OF LINES

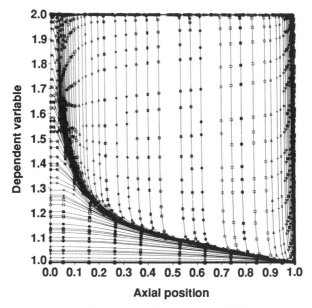

Figure 6.5.3: Single Step Reaction - All Time Steps

CASE	NODES	STEPS	FNS	JACS	ETF	CTF
1	32	624	1800	807	5	284
2	29	1125	3444	1544	9	528
3	32	614	1779	819	7	276
4	29	1183	3517	1728	559	12

Table 6.5.1: Integration Statistics for a Single Step Reaction

we have not discussed any of the implementation details required for an adaptive mesh code, such as how to select the initial mesh or how to add and delete points. We note that the filtered error estimate of Section 5.4.2 was found to be quite effective for this class of problems [204].

Gas Transmission Networks

We conclude this section with a brief discussion of an application where the utilization of a linear multistep method based on BDF, such as DASSL or LSODI, is not appropriate. This example illustrates that one must understand the application well in order to make a wise selection of a numerical method. In addition, since software for DAE's is still very much in the development stage, a thorough understanding of the differences between ODE's and DAE's is essential in writing a new DAE code. Critical design issues for a DAE code have been discussed in [134] as they pertain to the implementation of linear multistep methods. Similar issues must be considered in the design of any DAE code. A new DAE code was deemed essential to the efficient numerical solution of DAE's arising in the simulation of large scale gas transmission networks [73].

Let a sequence of pipes be connected to machines to form a large scale gas transmission network. Machines, which control the flow of the gas through the network, may be valves, sources, compressors, or regulators. Let the nodes be the points in the network where the machines and pipes are connected. A network is typically composed of at least fifty machines and 200 nodes. Two parabolic partial differential equations describing the momentum and continuity of the transient gas flow in the network are used to model this problem. After discretizing the spatial variable via the MOL, the problem is described by the following DAE

$$Ey' = A(y)y - d(t),$$

where y is a vector of unknown pressures at the nodes and gas flows through the machines. The vector $d(t)$ represents the consumer demands on the network, the matrix $A(y)$ depends on specific network properties (such as the conductivities of the pipes and operating characteristics of the machines), and the matrix E is determined by the pipe capacities and isothermal speed of sound. Generally these systems are very large and sparse since there is very little coupling in the network. A peculiarity of these problems is the fact that the demand vector $d(t)$ models large changes reflecting daily events of humans, and hence is represented as a piecewise continuous function. In addition, the starting and stopping of machines in the network, or simply the switching of their operating modes, introduce additional large disturbances which are reflected by rapid changes in the solution. An accurate numerical solution is not usually required during these periods of rapid change, but is of interest once the system has settled down. If a typical linear multistep (stiff) integrator such as DASSL or LSODI is applied, very small time steps are required during the transient regions and frequent restarts are initiated when the demand function or operating characteristics of the network switches. For the solution of the DAE's arising in the simulation of gas transmission networks, it is critical to design the numerical method and corresponding code to satisfy several criteria. For example, it must conserve the mass flow of the gas. It should be low order since typically only two significant figures in the solution are obtainable due to the limited known accuracy of the initial data and solution characteristics. Experience has shown that when the pipes in the network are relatively short, the DAE system is stiff. Finally, the code must include an error control mechanism that functions appropriately in both the rapidly and slowly changing regions of the solution. Chua and Dew designed a variable step integrator based on a Rosenbrock method having an imbedded local error estimator. For comparison, they also examined three other implicit

6.5. DAE'S ARISING FROM THE METHOD OF LINES

Runge-Kutta methods. Details of their code implementation and numerical results are given in [73].

In a similar application, but with fewer discontinuities, Winters [243] has designed the TOPAZ code to model the flow of fluid in pipes. To accomodate the possibility of choked flows, this application requires a root solver, as described in Section 5.3.3. Thus the TOPAZ code calls DASSLRT. A related problem which requires modeling the flow of fluid in pipes arises when modeling superconducting magnets [244]. This application also makes use of DASSLRT.

The need for a DAE solver to model combustion ignition and the flow of fluid in pipes for the modeling of solar central receiver power plants first motivated the writing of the DASSL code. Applications such as mechanical systems, electrical networks, trajectory prescribed path control and method of lines solution of partial differential equations have had an enormous impact on the development of both the numerical analysis and mathematical software for DAE systems, and will undoubtedly continue to influence the future developments in this area.

Bibliography

[1] S. Adjerid and J. E. Flaherty, *A moving finite element method with error estimation and refinement for one-dimensional time dependent partial differential equations*, SIAM J. Numer. Anal., 23 (1986), 778-796.

[2] R. Alexander, *Diagonally implicit Runge-Kutta methods for stiff ODEs*, SIAM J. Numer. Anal., 14 (1977), 1006-1022.

[3] R. Alexander and J. Coyle, *Runge-Kutta methods and differential-algebraic systems*, Iowa State University, Preprint, 1988.

[4] V. I. Arnold, *Mathematical Methods of Classical Mechanics*, Springer, Berlin, New York, 1978.

[5] U. Ascher, *On some difference schemes for singular singularly-perturbed boundary value problems*, Numer. Math., 46 (1986), 1-30.

[6] U. Ascher, *On numerical differential algebraic problems with application to semiconductor device simulation*, SIAM J. Numer. Anal., 26 (1989), 517-538.

[7] U. Ascher, *Symmetric schemes and differential algebraic equations*, SIAM J. Sci. Statist. Comput., 10 (1989), 937-949.

[8] U. Ascher and G. Bader, *Stability of collocation at Gaussian points*, SIAM J. Numer. Anal., 23 (1986), 412-422.

[9] U. Ascher and R. Weiss, *Collocation for singular perturbation problems I: First order systems with constant coefficients*, SIAM J. Numer. Anal., 20 (1983), 537-557.

[10] U. Ascher and R. Weiss, *Collocation for singular perturbation problems II: Linear first order systems without turning points*, Math. Comp., 43 (1984), 157-187.

[11] U. Ascher and R. Weiss, *Collocation for singular perturbation problems III: Nonlinear problems without turning points*, SIAM J. Sci. Statist. Comput., 5 (1984), 811-829.

[12] M. Athans and P. L. Falb, *Optimal Control*, McGraw-Hill, New York, 1966.

[13] G. Bader and P. Deuflhard, *A semi-implicit mid-point rule for stiff systems of ordinary differential equations*, Numer. Math., 41 (1983), 373-398.

[14] J. Baumgarte, *Stabilization of constraints and integrals of motion in dynamical systems*, Comput. Methods Appl. Mech. Engrg., 1 (1972), 1-16.

[15] M. Berzins, *A C^1 interpolant for codes based on backward difference formulae*, Appl. Numer. Math., 2 (1986), 109-118.

[16] M. Berzins, P. M. Dew, and R. M. Furzeland, *Developing PDE software using the method of lines and differential algebraic integrators*, Appl. Numer. Math., 5 (1989), 375-397.

[17] M. Berzins, P. M. Dew, and A. J. Preston, *Integration algorithms for the dynamic simulation of production processes*, Proc. First Europ. Conf. on Ind. Math., S. McKee, ed., 1989.

[18] M. Berzins and R. M. Furzeland, *A user's manual for SPRINT: Part 1*, Dept. of Computer Studies Report 199, Leeds University, Leeds, UK, 1985.

[19] M. Berzins and R. M. Furzeland, *A user's manual for SPRINT: Part 2*, Dept. of Computer Studies Report 202, Leeds University, Leeds, UK, 1986.

[20] M. Berzins, R. M. Furzeland, and P. M. Dew, *Software tools for time-dependent differential equations*, in Simulation and Optimization of Large Systems, A. J. Osiadacz, ed., Oxford University Press, Oxford, 1988.

[21] J. Betts, T. Bauer, K. Zondervan, and W. Huffman, *Solving the optimal control problem using a nonlinear programming technique*, AIAA/AAS Astrodynamics Conf., Seattle, WA, 1984.

[22] T. A. Bickart, *An efficient solution process for implicit Runge-Kutta methods*, SIAM J. Numer. Anal., 14 (1977), 1022-1027.

[23] L. T. Biegler, *On the simultaneous solution and optimization of large scale engineering systems*, Proc. XVIII Congress on the Use of Computers in Chemical Engineering, Giardini, Naxos, Itali, April 1987.

[24] L. T. Biegler, J. J. Damiano, and G. E. Blau, *Nonlinear parameter estimation: A case study comparison*, AIChE J., 32 (1986), 29-45.

[25] M. Bier, O. A. Palusinski, R. A. Mosher, and D. A. Saville, *Electrophoresis: Mathematical modeling and computer simulation*, Science, 219 (1983), 1281-1287.

[26] A. M. Bos, *Modeling Multibody Systems in Terms of Multibond Graphs with Application to a Motorcycle*, Ph.D. Thesis, Twente University, The Netherlands, 1986.

[27] R. K. Brayton, F. G. Gustavson, and G. D. Hachtel, *A new efficient algorithm for solving differential-algebraic systems using implicit backward differentiation formulas*, Proc. IEEE, 60 (1972), 98-108.

[28] W. G. Breiland, M. E. Coltrin, and P. Ho, *Comparisons between a gas-phase model of silane chemical vapor deposition and laser-diagnostic measurements*, J. Appl. Phys., 59 (1986), 3267-3273.

[29] W. G. Breiland, P. Ho, and M. E. Coltrin, *Gas-phase silicon atoms in silane chemical vapor deposition: Laser-excited fluorescence measurements and comparisons with model predictions*, J. Appl. Phys., 60 (1986), 1505-1513.

[30] K. E. Brenan, *Stability and Convergence of Difference Approximations for Higher Index Differential-Algebraic Systems with Applications in Trajectory Control*, Ph.D. Thesis, Univ. of California at Los Angeles, 1983.

[31] K. E. Brenan, *Numerical experiments with the backward difference method for a nonlinear differential-algebraic system, the pendulum problem*, Report No. 83-5, Dept. of Computer Science, Uppsala Univ., Uppsala, Sweden, 1983.

[32] K. E. Brenan, *Numerical solution of trajectory prescribed path control problems by backward differentiation formulas*, IEEE Trans. Automat. Control, AC-31 (1986), 266-269.

[33] K. E. Brenan and B. E. Engquist, *Backward differentiation approximations of nonlinear differential/algebraic equations*, and Supplement, Math. Comp., 51 (1988), 659-676, S7-S16.

[34] K. E. Brenan and L. R. Petzold, *The numerical solution of higher index differential/algebraic equations by implicit Runge-Kutta methods*, SIAM J. Numer. Anal., 26 (1989), 976-996.

[35] R. L. Brown and C. W. Gear, *Documentation for DFASUB – A program for the solution of simultaneous implicit differential and nonlinear equations*, Technical Report UIUCDCS-R-73-575, University of Illinois at Urbana-Champaign, 1973.

[36] P. N. Brown and A. C. Hindmarsh, *Reduced storage matrix methods in stiff ODE systems*, Appl. Math. Comput., 31 (1989), 40-91.

[37] A. E. Bryson and Y. C. Ho, *Applied Optimal Control*, Hemisphere Publishing Corp., Washington, DC, 1975.

[38] K. Burrage, *A special family of Runge-Kutta methods for solving stiff differential equations*, BIT, 18 (1978), 22-41.

[39] K. Burrage, *Efficiently implementable algebraically stable Runge-Kutta methods*, SIAM J. Numer. Anal., 19 (1982), 245-258.

[40] K. Burrage, J. C. Butcher, and F. H. Chipman, *An implementation of singly-implicit Runge-Kutta methods*, BIT, 20 (1980), 326-340.

[41] K. Burrage and W. H. Hundsdorfer, *The order of B-convergence of algebraically stable Runge-Kutta methods*, BIT, 27 (1987), 62-71.

[42] K. Burrage and L. Petzold, *On order reduction for Runge-Kutta methods applied to differential/algebraic systems and to stiff systems of ODEs*, SIAM J. Numer. Anal., 27 (1990), 447-456.

[43] K. Burrage, W. Hundsdorfer, and J. G. Verwer, *A study of B-convergence of Runge-Kutta methods*, Computing, 36 (1986), 17-34.

[44] J. C. Butcher, *Implicit Runge-Kutta processes*, Math. Comp., 18 (1964), 50-64.

[45] J. C. Butcher, *On the implementation of Runge-Kutta methods*, BIT, 16 (1976), 237-240.

[46] G. D. Byrne and P. R. Ponzi, *Differential-algebraic systems, their applications and solutions*, Computers and Chemical Engineering, 12 (1988), 377-382.

[47] M. Calvo, F. Lisbona, and J. Montijano, *On the stability of variable step-size Nordsieck BDF methods*, SIAM J. Numer. Anal., 24 (1987), 844-854.

[48] I. T. Cameron, *Solution of differential-algebraic systems using diagonally implicit Runge-Kutta methods*, IMA J. Numer. Anal., 3 (1983), 273-289.

[49] S. L. Campbell, *Singular Systems of Differential Equations*, Pitman, Boston, 1980.

[50] S. L. Campbell, *Singular Systems of Differential Equations II*, Pitman, Boston, 1982.

[51] S. L. Campbell, *Multiparameter singular systems: Nondissective approaches*, Proc. 22nd IEEE Conference on Decision and Control, 1983, 1278-1281.

[52] S. L. Campbell, *One canonical form for higher index linear time varying singular systems*, Circuits Systems Signal Process., 2 (1983), 311-326.

[53] S. L. Campbell, *Regularizations of linear time varying singular systems*, Automatica, 20 (1984), 365-370.

[54] S. L. Campbell, *The numerical solution of higher index linear time-varying singular systems of differential equations*, SIAM J. Sci. Statist. Comput., 6 (1985), 334-338.

[55] S. L. Campbell, *Rank deficient least squares and the numerical solution of linear singular implicit systems of differential equations*, Contemp. Math., 47 (1985), 51-63.

[56] S. L. Campbell, *Index two linear time varying singular systems of differential equations*, Circuits Systems Signal Process., 5 (1986), 97-108.

[57] S. L. Campbell, *Consistent initial conditions for linear time varying singular systems*, in Frequency Domain and State Space Methods, C. I. Byrnes and A. Lindquist, eds., North Holland, Amsterdam, 1986, 313-318.

[58] S. L. Campbell, *A general form for solvable linear time varying singular systems of differential equations*, SIAM J. Math. Anal., 18 (1987), 1101-1115.

[59] S. L. Campbell, *A computational method for general higher index singular systems of differential equations*, IMACS Transactions Scientific Computing-'88, 1.2 (1989), 555-560.

[60] S. L. Campbell, *Control problem structure and the numerical solution of linear singular systems*, Math. Control Signals Systems, 1 (1988), 73-87.

[61] S. L. Campbell, *Distributional convergence of BDF approximations to descriptor systems*, Circuits Systems Signal Process., 8 (1989), 261-265.

[62] S. L. Campbell and K. D. Clark, *Order and the index of singular time-invariant linear systems*, Systems Control Lett., 1 (1981), 119-122.

[63] S. L. Campbell and K. D. Clark, *Singular control problem structure and the convergence of backward differentiation formulas*, in Linear Circuits, Systems and Signal Processing: Theory and Application, C. I. Brynes, C. F. Martin, and R. E. Saeks, eds., North-Holland, Amsterdam, 1988, 19-26.

[64] S. L. Campbell and K. D. Clark, *Convergence of BDF approximations for nonsolvable differential algebraic equations*, Appl. Numer. Math., 6 (1989/1990), 153-158.

[65] S. L. Campbell and L. R. Petzold, *Canonical forms and solvable singular systems of differential equations*, SIAM J. Alg. Discrete Methods, 4 (1983), 517-521.

[66] M. Caracotsios and W. E. Stewart, *Sensitivity analysis of initial value problems with mixed ODEs and algebraic equations*, Computers and Chemical Engineering, 9, (1985), 359-365.

[67] J. R. Cash, *Diagonally implicit Runge-Kutta formulae with error estimates*, J. Inst. Math. Appl., 24 (1979), 293-301.

[68] F. H. Chipman, *A-stable Runge-Kutta processes*, BIT, 11 (1971), 384-388.

[69] J. C. K. Chou, G. Baciu, and H. K. Kesavan, *Graph-theoretic models for simulating robot manipulators*, Proc. 1987 IEEE International Conf. on Robotics and Automation, Raleigh, NC, 2 (1987), 953-959.

[70] J. C. K. Chou, G. Baciu, and H. K. Kesavan, *Computational scheme for simulating robot manipulators*, Proc. 1987 IEEE International Conf. Robotics and Automation, Raleigh, NC, 2 (1987), 961-966.

[71] L. O. Chua and P. M. Lin, *Computer-Aided Analysis of Electronic Circuits*, Prentice-Hall, Englewood Cliffs, NJ, 1975.

[72] L. O. Chua and R. A. Rohrer, *On the dynamic equations of a class of nonlinear RLC networks*, IEEE Trans. Circuit Theory, CT-12 (1965), 475-489.

[73] T. S. Chua and P. M. Dew, *The design of a variable-step integrator for the simulation of gas transmission networks*, Int. J. Numer. Methods Engrg., 20, (1984), 1797-1813.

[74] K. D. Clark, *Difference methods for the numerical solution of time varying singular systems of differential equations*, SIAM J. Alg. Discrete Methods, 7 (1986), 236-246.

[75] K. D. Clark, *The numerical solution of some higher index time varying semistate systems by difference methods*, Circuits Systems Signal Process., 6 (1987), 61-75.

[76] K. D. Clark, *A structural form for higher index semistate equations I: Theory and applications to circuit and control*, Linear Algebra Appl., 98 (1988), 169-197.

[77] K. D. Clark and L. R. Petzold, *Numerical solution of boundary value problems in differential/algebraic systems*, SIAM J. Sci. Statist. Comput., 10 (1989), 915-936.

[78] J. D. Cobb, *On the solutions of linear differential equations with singular coefficients*, J. Differential Equations, 46 (1982), 310-323.

[79] J. D. Cobb, *A further interpretation of inconsistent initial conditions in descriptor-variable systems*, IEEE Trans. Automat. Control, AC-28 (1983), 920-922.

[80] M. E. Coltrin, R. J. Kee, and J. A. Miller, *A mathematical model of the coupled fluid mechanics and chemical kinetics in a chemical vapor deposition reactor*, J. Electrochem. Soc., 131 (1984), 425-434.

[81] M. E. Coltrin, R. J. Kee, and J. A. Miller, *A mathematical model of silicon chemical vapor deposition*, J. Electrochem. Soc., 133 (1986), 1206-1213.

[82] M. Crouzeix and F. Lisbona, *The convergence of variable-stepsize, variable-formula multistep methods*, SIAM J. Numer. Anal., 21 (1984), 512-534.

[83] P. A. Cundall, *Rational design of tunnel support*, Dept. of Civil and Mining Engineering, Univ. of Minnesota, Minneapolis, MN, 1974.

[84] A. R. Curtis, *The FACSIMILE numerical integrator for stiff initial value problems*, in Computational Techniques for Ordinary Differential Equations, I. Gladwell and D. K. Sayers, eds., Academic Press, New York, 1980, 47-82.

[85] A. R. Curtis, M. J. D. Powell, and J. K. Reid, *On the estimation of sparse Jacobian matrices*, J. Inst. Math. Appl., 13 (1974), 117-120.

[86] G. Dahlquist and A. Björk, *Numerical Methods,* N. Anderson, translator, Prentice-Hall, Englewood Cliffs, NJ, 1974.

[87] G. Dahlquist and B. Lindberg, *On some implicit one-step methods for stiff differential equations*, Technical Report TRITA-NA-7302, Stockholm, 1973.

[88] A. De Luca, *Control properties of robot arms with joint elasticity*, Proc. Math. Theory Networks and Systems, 1987.

[89] A. De Luca and A. Isidori, *Feedback linearization of invertible systems*, Colloquium on Automation and Robotics, Duisburg, July 1987.

[90] A. De Luca, A. Isidori, and F. Nicolo', *An application of nonlinear model matching to the dynamic control of robot arm with elastic joints*, Proc. IFAC Symposium on Robot Control, Barcelona, 1985, 55-62.

[91] A. De Luca, A. Isidori, and F. Nicolo', *Control of robot arm with elastic joints via nonlinear dynamic feedback*, Proc. 24th IEEE Conf. Decision and Control, 1985, 1671-1679.

[92] K. Dekker and E. Hairer, *A necessary condition for BSI-stability*, BIT, 25 (1985), 285-288.

[93] K. Dekker and J. G. Verwer, *Stability of Runge-Kutta methods for stiff nonlinear differential equations*, Elsevier Science Publishers, New York, 1984.

[94] P. Deuflhard, *Order and step-size control in extrapolation methods*, Numer. Math., 41 (1983), 399-422.

[95] P. Deuflhard, *Recent progress in extrapolation methods for ordinary equations*, SIAM Rev., 27 (1985), 273-289.

[96] P. Deuflhard, E. Hairer, and J. Zugck, *One step and extrapolation methods for differential-algebraic systems*, Numer. Math., 51 (1987), 501-516.

[97] P. Deuflhard and U. Nowak, *Extrapolation integrators for quasilinear implicit ODEs*, in Large Scale Scientific Computing, P. Deuflhard and B. Engquist, eds., Progress in Scientific Computing 7, Birkhäuser, Basel, 1987, 37-50.

[98] E. D. Dickmanns, *Maximum range three-dimensional lifting planetary entry*, Technical Report NASA-TR-R-387, August 1972.

[99] V. Dolezal, *The existence of a continuous basis of a certain linear subspace of E_r, which depends on a parameter*, Časopis pro pěstováni matematiky, roč. 89, Praha, 1964, 466-468.

[100] J. J. Dongarra, J. R. Bunch, C. B. Moler, and G. W. Stewart, *LINPACK User's Guide*, SIAM, Philadelphia, 1979.

[101] I. Duff and C. W. Gear, *Computing the structural index*, SIAM J. Alg. Discrete Methods, 7 (1986), 594-603.

[102] B. Dziurla and R. W. Newcomb, *Parasitics in the semistate formulation*, Proc. 26th Midwest Symp. Circuits Systems, 1983, 578-580..

[103] R. Frank, J. Schneid, and C. W. Ueberhuber, *The concept of B-convergence*, SIAM J. Numer. Anal., 18 (1981), 753-780.

[104] R. Frank, J. Schneid, and C. W. Ueberhuber, *B-convergence of Runge-Kutta methods*, Bericht Nr. 48/81, Institut für Numerische Mathematik, TU-Wien, 1981.

[105] R. Frank, J. Schneid, and C. W. Ueberhuber, *Stability properties of implicit Runge-Kutta methods*, SIAM J. Numer. Anal., 22 (1985), 497-514.

[106] R. Frank, J. Schneid, and C. W. Ueberhuber, *Order results for implicit Runge-Kutta methods applied to stiff systems*, SIAM J. Numer. Anal., 22 (1985), 515-534.

[107] C. Führer, *Differential-Algebraische Gleichungsysteme in Mechanischen Mehrkörpersystemen*, Ph.D. Thesis, Math. Institute, Technische Universität, München, 1988.

[108] C. Führer and B. J. Leimkuhler, *Numerical solution of differential-algebraic equations for constrained mechanical motion*, Numer. Math., 59 (1991), 55-69.

[109] C. Führer and O. Wallrapp, *A computer-oriented method for reducing linearized multibody equations by incorporating constraints*, Comput. Methods Appl. Mech. Engrg., 46 (1984), 169-175.

[110] R. Gani, C. A. Ruiz, and I. T. Cameron, *A generalized model for distillation columns*, Comput. Chem. Engrg., 10 (1986), 181-198.

[111] F. R. Gantmacher, *The Theory of Matrices*, Chelsea, New York, 1959.

[112] F. R. Gantmacher, *The Theory of Matrices*, Vol. 2, Chelsea, New York, 1964.

[113] C. W. Gear, *The simultaneous numerical solution of differential-algebraic equations*, IEEE Trans. Circuit Theory, CT-18 (1971), 89-95.

[114] C. W. Gear, *Numerical Initial Value Problems in Ordinary Differential Equations*, Prentice-Hall, Englewood Cliffs, NJ, 1971.

[115] C. W. Gear, *Asymptotic estimation of errors and derivatives for the numerical solution of ordinary differential equations*, IFIP 74, 447-451.

[116] C. W. Gear, *Method and initial stepsize selection in multistep ODE solvers*, Technical Report UIUCDCS-R-80-1006, Dept. of Computer Science, Univ. of Illinois, Urbana, IL, 1980.

[117] C. W. Gear, *Maintaining solution invariants in the numerical solution of ODEs*, SIAM J. Sci. Statist. Comput., 7 (1986), 734-743.

[118] C. W. Gear, *Differential-algebraic equation index transformations*, SIAM J. Sci. Statist. Comput., 9 (1988), 39-47.

[119] C. W. Gear, H. H. Hsu, and L. Petzold, *Differential-algebraic equations revisited*, Proc. ODE Meeting, Oberwolfach, West Germany, 1981.

[120] C. W. Gear and J. Keiper, personal correspondence to L. Petzold, 1989.

[121] C. W. Gear, B. Leimkuhler, and G. K. Gupta, *Automatic integration of Euler-Lagrange equations with constraints*, J. Comput. Appl. Math., 12/13 (1985), 77-90.

[122] C. W. Gear and L. R. Petzold, *Differential/algebraic systems and matrix pencils*, in Matrix Pencils, B. Kagstrom and A. Ruhe, eds., Lecture Notes in Mathematics 973, Springer-Verlag, Berlin, New York, 1983, 75-89.

[123] C. W. Gear and L. R. Petzold, *ODE methods for the solution of differential/algebraic systems*, SIAM J. Numer. Anal., 21 (1984), 367-384.

[124] C. W. Gear and K. W. Tu, *The effect of variable mesh size on the stability of multistep methods*, SIAM J. Numer. Anal., 11 (1974), 1025-1043.

[125] C. W. Gear and D. S. Watanabe, *Stability and convergence of variable order multistep methods*, SIAM J. Numer. Anal., 11 (1974), 1044-1058.

[126] G. H. Golub and C. F. Van Loan, *Matrix Computations*, Johns Hopkins University Press, Baltimore, 1983.

[127] W. B. Gragg, *On extrapolation algorithms for ordinary initial value problems*, SIAM J. Numer. Anal., 2 (1965), 384-404.

[128] P. M. Gresho, S. T. Chan, R. L. Lee, and C. D. Upson, *A modified finite element method for solving the time-dependent incompressible Navier-Stokes equations. Part 1: Theory*, Internat. J. Numer. Methods Fluids, 4 (1984), 557-598.

[129] P. M. Gresho and C. D. Upson, *Current progress in solving the time-dependent, incompressible Navier-Stokes equations in three-dimensions by (almost) the FEM*, Proc. Fourth Int. Conf. on Finite Elements in Water Resources, Springer-Verlag, Berlin, New York, 1982, 43-58.

[130] E. Griepentrog, *Initial value problems for implicit differential equation systems*, in Numerical Methods for Ordinary Boundary Value Problems, R. März, ed., Seminarbericht Nr. 55, Humboldt Univ., Berlin, 1984, 21-48.

[131] E. Griepentrog and C. Bandt, *Transferability of higher-index differential-algebraic equations*, Preprint 20, Ernst Moritz Arndt Universität Greifswald, Sektion Mathematik, 1988.

[132] E. Griepentrog and R. März, *Differential-Algebraic Equations and Their Numerical Treatment*, Teubner-Texte zur Mathematik, Band 88, 1986.

[133] R. D. Grigorieff, *Stability of multistep methods on variable grids*, Numer. Math., 42 (1983), 359-377.

[134] G. K. Gupta, C. W. Gear, and B. Leimkuhler, *Implementing linear multistep formulas for solving DAEs*, Technical Report UIUCDCS-R-85-1205, Dept. of Computer Science, Univ. of Illinois at Urbana-Champaign, April 1985.

[135] E. Hairer and C. Lubich, *Extrapolation at stiff differential equations*, Numer. Math., 52 (1988), 377-400.

[136] E. Hairer, C. Lubich, and M. Roche, *The numerical solution of differential-algebraic systems by Runge-Kutta methods*, Springer-Verlag, Berlin, 1989.

[137] G. Hall and J. M. Watt, *Modern Numerical Methods for Ordinary Differential Equations*, Clarendon Press, Oxford, 1976.

[138] P. C. Hammer and J. W. Hollingsworth, *Trapezoidal methods of approximating solutions of differential equations*, M.T.A.C., 9 (1955), 92-96.

[139] M. Hanke, *Linear differential-algebraic equations in spaces of integrable functions*, Preprint 114, Humboldt-Universität zu Berlin, Sektion Mathematik, 1986.

[140] M. Hanke, *On the regularization of index 2 differential-algebraic equations*, Preprint 137, Humboldt-Universität zu Berlin, Sektion Mathematik, 1987.

[141] A. Harten and J. M. Hyman, *A self-adjusting grid for the computation of weak solutions of hyperbolic conservation laws*, J. Comput. Phys., 50 (1983), 235-269.

[142] K. L. Hiebert and L. F. Shampine, *Implicitly defined output points for solutions of ODEs*, Technical Report SAND 80-0180, Sandia National Laboratories, Albuquerque, NM, 1980.

[143] A. C. Hindmarsh, *LSODE and LSODI, two new initial value ordinary differential equation solvers*, ACM-SIGNUM Newsletters, 15 (1980), 10-11.

[144] A. C. Hindmarsh, *ODEPACK, A systematized collection of ODE solvers*, in Scientific Computing, R. S. Stepleman et al., eds., North-Holland, Amsterdam, 1983, 55-64.

[145] N. Houbak and P. G. Thomsen, *A FORTRAN subroutine for the solution of stiff ODEs with sparse Jacobians*, Technical University of Denmark, ISSN 0105-4988, 1979.

[146] J. M. Hyman, *Moving mesh methods for partial differential equations*, in Mathematics Applied to Science, J. Goldstein, S. Rosencrans, and G. Sod, eds., Academic Press, New York, 1988, 129-154.

[147] K. R. Jackson and R. Sacks-Davis, *An alternative implementation of variable step-size multistep formulas for stiff ODEs*, ACM Trans. Math. Software, 6 (1980), 295-318.

[148] S. H. Johnson and A. C. Hindmarsh, *MSRS: A Fortran code for the numerical dynamic simulation of solid/fluid reactions in non-isothermal multispecies porous spheres with Stefan-Maxwell diffusion*, UCRL-21022, Lawrence Livermore National Laboratory, 1988.

[149] B. Kagstrom and A. Ruhe, editors, *Matrix Pencils*, Vol. 973, Springer-Verlag, Berlin, New York, 1983.

[150] E. J. Kansa, D. L. Morgan, and L. K. Morris, *A simplified moving finite difference scheme: Application to dense gas dispersion*, SIAM J. Sci. Statist. Comput., 5 (1984), 667-683.

[151] R. J. Kee and L. R. Petzold, *A differential/algebraic equation formulation of the method-of-lines solution to systems of partial differential equations*, Technical Report SAND86-8893, Sandia National Laboratories, Albuquerque, NM, 1986.

[152] R. J. Kee, L. R. Petzold, M. D. Smooke, and J. F. Grcar, *Implicit methods in combustion and chemical kinetics modeling*, in Multiple Time Scales, Academic Press, New York, 1985.

[153] M. Knorrenschild, *Regularisierungen von Differentiell-Algebraischen Systemen—Theoretische und Numerische Aspekte*, Ph.D. Thesis, RWTH Aachen, 1988.

[154] J. Kraaijevanger, *B-convergence of the implicit midpoint rule and the trapezoidal rule*, BIT, 25 (1985), 652-666.

[155] F.T. Krogh, *Changing step size in the integration of differential equations using modified divided differences*, Proc. Conf. Num. Solution of ODEs, Lecture Notes in Mathematics 362, Springer-Verlag, New York, 1974.

[156] F. T. Krogh and K. Stewart, *Asymptotic ($h \to \infty$) absolute stability for BDFs applied to stiff differential equations*, ACM Trans. Math. Software, 10 (1984), 45-57.

[157] A. Kværno, *Order conditions for Runge-Kutta methods applied to differential-algebraic systems of index one*, Technical Report, Div. of Mathematical Sciences, Univ. of Trondheim, Norway, 1987.

[158] A. Kværno, *Runge-Kutta methods applied to fully implicit differential/algebraic equations of index one*, Math. Comp., 54 (1990), 583-625.

[159] J. D. Lambert, *Computational Methods in Ordinary Differential Equations*, Wiley, New York, 1973.

[160] B. J. Leimkuhler, *Error estimates for differential-algebraic equations*, Technical Report UIUCDCD-R-86-1287, Dept. of Computer Science, Univ. of Illinois, 1986.

[161] B. J. Leimkuhler, *Approximation Methods for the Consistent Initialization of Differential-Algebraic Equations*, Technical Report UIUCDCS-R-88-1450 (Ph.D. Thesis), Dept. of Computer Science, Univ. of Illinois, 1988.

[162] B. J. Leimkuhler, L. R. Petzold, and C. W. Gear, *Approximation methods for the consistent initialization of differential-algebraic systems of equations*, SIAM J. Numer. Anal., 28 (1991), 205-226.

[163] F. L. Lewis, *A survey of linear singular systems*, Circuits Systems Signal Process., 5 (1986), 3-36.

[164] W. Liniger, *Multistep and one-leg methods for implicit mixed differential algebraic systems*, IEEE Trans. Circuits Systems, CAS-26 (1979), 755-762.

[165] P. Lötstedt, *On a penalty function method for the simulation of mechanical systems subject to constraints*, Technical Report TRITA-NA-7919, Royal Institute of Technology, Stockholm, Sweden, 1979.

[166] P. Lötstedt, *Mechanical systems of rigid bodies subject to unilateral constraints*, SIAM J. Appl. Math., 42 (1982), 281-296.

[167] P. Lötstedt, *Discretization of singular perturbation problems by BDF methods*, Report No. 99, Uppsala Univ., Dept. of Computer Science, 1985.

[168] P. Lötstedt, *On the relation between singular perturbation problems and differential-algebraic equations*, Report No. 100, Uppsala Univ., Dept. of Computer Science, 1985.

[169] P. Lötstedt and G. Dahlquist, *Interactive simulation of the progressive collapse of a house*, in Numerical Methods for Differential Equations and Simulation, A. W. Bennet and R. Vichnevetsky, eds., North-Holland, New York, 1978, 135-138.

[170] P. Lötstedt and L. Petzold, *Numerical solution of nonlinear differential equations with algebraic constraints I: Convergence results for backward differentiation formulas*, Math. Comp., 46 (1986), 491-516.

[171] Ch. Lubich, *Extrapolation methods for differential-algebraic systems*, Technical Report, Institut für Mathematik und Geometrie, Universität Innsbruck, Austria, 1988.

[172] Ch. Lubich, h^2 - *extrapolation methods for differential-algebraic systems of index 2*, Impact. Comput. Sci. Engrg., 1 (1989), 260-268.

[173] Ch. Lubich and M. Roche, *Rosenbrock methods for differential-algebraic systems with solution-dependent singular matrix multiplying the derivative*, Comput. Arch. Inf. Numer., 43 (1990), 325-342.

[174] D. G. Luenberger, *Dynamic equations in descriptor form*, IEEE Trans. Automat. Control, AC-22 (1977), 312-321.

[175] D. G. Luenberger, *Time invariant descriptor systems*, 1977 Joint Automatic Control Conference, San Francisco, CA, 725-730.

[176] A. E. Lutz, R. J. Kee, and H. A. Dwyer, *Ignition modeling with grid adaption*, Proc. 10th International Colloquium on Gas Dynamics of Explosions and Reactive Systems, Berkeley, CA, 1980.

[177] U. Maas and J. Warnatz, *Ignition processes in carbon-monoxide-hydrogen-oxygen mixtures*, Proc. 22nd Int. Symp. on Combustion, Seattle, WA, 1988.

[178] I. Mack, *Block Implicit One-Step Methods for Solving Smooth and Discontinuous Systems of Differential/Algebraic Equations*, Ph.D. Thesis, Harvard Univ., 1986.

[179] J. W. Manke, B. Dembart, M. A. Epton, A. M. Erisman, Paul Lu, R. F. Sincovec, and E. L. Yip, *Solvability of large scale descriptor systems*, Report, Boeing Computer Services Company, Seattle, WA, 1979.

[180] W. Marquardt, Universität Stuttgart, personal communication to L. Petzold, 1988.

[181] R. März, *On difference and shooting methods for boundary value problems in differential-algebraic equations*, Preprint Nr. 24, Humboldt-Universität zu Berlin, 1982.

[182] R. März, *On initial value problems in differential-algebraic equations and their numerical treatment*, Computing, 35 (1985), 13-37.

[183] R. März, *On tractability with index 2*, Preprint 109, Humboldt-Universität zu Berlin, 1986.

[184] R. März, *A matrix chain for analyzing differential-algebraic equations*, Preprint 162, Humboldt-Universität zu Berlin, Sektion Mathematik, 1987.

[185] R. März, *Analysis and numerical treatment of differential-algebraic systems*, Proc. Oberwolfach-Konferenz Mathematische Modellierung und Simulation Elektrischer Schaltungen, 1988.

[186] R. März, *Index-2 Differential-Algebraic Equations*, Results Math., 15 (1989), 149-171.

[187] N. H. McClamroch, *Singular systems of differential equations as dynamic models for constrained robot systems*, Technical Report RSD-TR-2-86, Univ. of Michigan Robot Systems Division, 1986.

[188] N. H. McClamroch and Han-Pang Huang, *Dynamics of a closed chain manipulator*, Amer. Control Conf. 50-4, 1, 1985.

[189] L. Meirovitch, *Methods of Analytical Dynamics*, McGraw-Hill, New York, 1970.

[190] T. Mrziglod, *Zur Theorie und Numerischen Realisierung von Lösungsmethodan bei Differential Gleichungen mit Angekoppelten Algebraischen Gleichungen,* Diplomarbeit, Universität zu Köln, 1987.

[191] R. W. Newcomb, *The semistate description of nonlinear and time-variable circuits*, IEEE Trans. Circuits Systems, CAS-26 (1981), 62-71.

[192] R. W. Newcomb, *Semistate design theory: Binary and swept hysteresis*, Circuits Systems Signal Process., 1 (1982), 203-216.

[193] P. E. Nikravesh, *Some methods for dynamical analysis of constrained mechanical systems: A survey*, in Computer Aided Analysis and Optimization of Mechanical System Dynamics, E. Haug, ed., Springer, Berlin, 1984, 223-259.

[194] Nørsett, S.P., *Runge-Kutta methods with a multiple real eigenvalue only*, BIT, 16 (1976), 388-393.

[195] R. E. O'Malley, *On nonlinear singularly perturbed initial value problems*, SIAM Rev., 30 (1988), 193-212.

[196] R. E. O'Malley and J. E. Flaherty, *Analytical and numerical methods for nonlinear singular singularly-perturbed initial value problems*, SIAM J. Appl. Math., 38 (1980), 225-248.

[197] N. Orlandea, D. A. Calahan, and M. A. Chace, *A sparsity-oriented approach to the dynamic analysis and design of mechanical systems: Part 1 and Part 2*, Trans. ASME J. Engrg. Ind., Ser. B, 99 (1977), 773-784.

[198] J. F. Painter, *Solving the Navier-Stokes equations with LSODI and the method of lines*, Report UCID-19262, Lawrence Livermore National Laboratory, 1981.

[199] C. C. Pantelides, *The consistent initialization of differential-algebraic systems*, SIAM J. Sci. Statist. Comput., 9 (1988), 213-231.

[200] L. Petzold, *Differential/algebraic equations are not ODEs*, SIAM J. Sci. Statist. Comput., 3 (1982), 367-384.

[201] L. R. Petzold, *A Description of DASSL: A differential/algebraic system solver*, in Scientific Computing, R. S. Stepleman et al., eds., North-Holland, Amsterdam, 1983, 65-68.

[202] L. R. Petzold, *Order results for implicit Runge-Kutta methods applied to differential/algebraic systems*, SIAM J. Numer. Anal., 23 (1986), 837-852.

[203] L. R. Petzold, *Observations on an adaptive moving grid method for one-dimensional systems of partial differential equations*, Appl. Numer. Math., 3 (1987), 347-360.

[204] L. R. Petzold, *An adaptive moving grid method for one-dimensional systems of partial differential equations and its numerical solution*, Proc. Workshop on Adaptive Methods for Partial Differential Equations, Rensselaer Polytechnic Institute, October 1988.

[205] L. R. Petzold and P. Lötstedt, *Numerical solution of nonlinear differential equations with algebraic constraints II: Practical implications*, SIAM J. Sci. Statist. Comput., 7 (1986), 720-733.

[206] A. Prothero and A. Robinson, *On the stability and accuracy of one-step methods for solving stiff systems of ordinary differential equations*, Math. Comp., 28 (1974), 145-162.

[207] F. Quareni and D. A. Yuen, *Time-dependent solutions of mean-field equations with applications for mantle convection*, Physics of the Earth and Planetary Interior, 36 (1984), 337-353.

[208] B. Raffel, J. Warnatz, H. Wolff, J. Wolfrum, and R. J. Kee, *Thermal ignition and minimum ignition energy on oxygen-ozone mixtures*, Proc. 10th International Colloquium on Gas Dynamics of Explosions and Reactive Systems, Berkeley, CA, 1985.

[209] S. Reich, *Differential-Algebraic Equations and Vector Fields on Manifolds*, Ph.D. Thesis, Technische Universität Dresden, 1988.

[210] P. Rentrop, M. Roche, and G. Steinebach, *The application of Rosenbrock-Wanner type methods with stepsize control in differential-algebraic equations*, Preprint TUM-M8804, Technische Universität München, 1988.

[211] W. C. Rheinboldt, *Differential-algebraic systems as differential equations on manifolds*, Math. Comp., 43 (1984), 473-482.

[212] R. E. Roberson and R. Schwertassek, *Dynamics of Multibody Systems*, Springer-Verlag, Heidelberg, 1988.

[213] M. Roche, *Rosenbrock methods for differential algebraic equations*, Numer. Math., 52 (1988), 45-63.

[214] M. Roche, *Implicit Runge-Kutta methods for differential algebraic equations*, SIAM J. Numer. Anal., 26 (1989), 963-975.

[215] M. Roche, *Méthods de Runge-Kutta et Rosenbrock pour Équations Différentielles Algëbriques et Systèmes Différentiels Raides*, Ph.D. Thesis, Université de Genève, 1988.

[216] T. Rübner-Peterson, *An efficient algorithm using backward time-scaled differences for solving stiff differential algebraic systems*, Institute of Circuit Theory and Telecommunication, Technical University of Denmark, 2800 Lyngby, 1973.

[217] R. Sacks-Davis, *Error estimates for a stiff differential equation procedure*, Math. Comp., 31 (1977), 939-953.

[218] D. E. Salane, *Adaptive routines for forming Jacobians numerically*, Technical Report SAND86-1319, Sandia National Laboratories, Albuquerque, NM, 1986.

[219] R. L. Sani, B. E. Eaton, P. M. Gresho, R. L. Le, and S. T. Chan, *On the solution of the time-dependent incompressible Navier-Stokes equations via a penalty Galerkin finite element method*, Technical Report UCRL-85354, Lawrence Livermore National Laboratory, 1981.

[220] R. E. Scraton, *Some L-stable methods for stiff differential equations*, Internat. J. Comput. Math., Section B, 9 (1981), 81-87.

[221] L. F. Shampine, *Implementation of implicit formulas for the solution of ODEs*, SIAM J. Sci. Statist. Comput., 1 (1980), 103-118.

[222] L. F. Shampine, *Conditioning of matrices arising in the solution of stiff ODEs*, Technical Report SAND82-0906, Sandia National Laboratories, Albuquerque, NM, 1982.

[223] L. F. Shampine, *Conservation laws and the numerical solution of ODEs*, Comput. Math. Appl., Part B, 12, (1986), 1287-1296.

[224] L. F. Shampine and L. S. Baca, *Error estimators for stiff differential equations*, J. Comput. Appl. Math., 11 (1984), 197-207.

[225] L. F. Shampine and M. K. Gordon, *Computer Solution of Ordinary Differential Equations*, W.H. Freeman and Co., San Francisco, CA, 1975.

[226] L. F. Shampine and H. A. Watts, *DEPAC – Design of a user oriented package of ODE solvers*, Technical Report SAND79-2374, Sandia National Laboratories, Albuquerque, NM, 1980.

[227] L. M. Silverman, *Inversion of multivariable systems*, IEEE Trans. Automat. Control, AC-14 (1969), 270-276.

[228] R. F. Sincovec, B. Dembart, M. A. Epton, A. M. Erisman, J. W. Manke, and E. L. Yip, *Solvability of large scale descriptor systems*, Technical Report, Boeing Computer Services Company, Seattle, WA, June 1979.

[229] R. F. Sincovec, A. M. Erisman, E. L. Yip, and M. A. Epton, *Analysis of descriptor systems using numerical algorithms*, IEEE Trans. Automat. Control, AC-26 (1981), 139-147.

[230] R. J. Skeel, University of Illinois, private communication, 1988.

[231] S. Skelboe, *The control of order and steplength for backward differentiation methods*, BIT, 17 (1977), 91-107.

[232] A. Skjellum, S. Mattisson, M. Morari, and L. Peterson, *Concurrent DASSL: Structure, Application, and Performance*, Proc. Fourth Conf. on Hypercube Concurrent Computers and Applications, Monterey, CA, 1989.

[233] G. Söderlind, *DASP3 - A program for the numerical integration of partitioned stiff ODE's and differential-algebraic systems*, Technical Report TRITA-NA-8008, Dept. of Numerical Analysis and Computing Science, Royal Institute of Technology, Sweden, 1980.

[234] J. W. Starner, *A Numerical Algorithm for the Solution of Implicit Algebraic-Differential Systems of Equations*, Ph.D. Thesis, Dept. of Mathematics and Statistics, Univ. of New Mexico, 1976.

[235] A. Van Der Sluis, *Condition, equilibration and pivoting in linear algebraic systems,* Numer. Math., 15 (1970), 74-86.

[236] J. M. Varah, *Stiff stability considerations for implicit Runge-Kutta methods*, Technical Report 80-1, Dept. of Computer Science, Univ. of British Columbia, Vancouver, Canada.

[237] H. A. Watts, *Starting step size for an ODE solver*, J. Comput. Appl. Math., 9 (1983), 177-191.

[238] H. A. Watts, *A smooth output interpolation process for BDF codes*, J. Appl. Math. Comp., 31 (1989), 397-418.

[239] H. A. Watts, *HSTART - an improved initial step size routine for ODE codes*, Technical Report SAND86-2633, Sandia National Laboratories, Albuquerque, NM, 1986.

[240] H. A. Watts and L. F. Shampine, *A-stable block implicit one-step methods*, BIT, 12 (1972), 252-266.

[241] R. A. Wehage and E. J. Haug, *Generalized coordinate partitioning for dimension reduction in analysis of constrained dynamic systems*, J. Mech. Design 134, (1982), 247-255.

[242] J. H. Wilkinson, *The practical significance of the Drazin inverse*, in Recent Applications of Generalized Inverses, S. L. Campbell, ed., Pitman, Boston, 1982, 82-99.

[243] W. S. Winters, *TOPAZ - The transient one-dimensional pipe flow analyzer: An update on code improvements and increased capabilities,* Technical Report SAND87-8225, Sandia National Laboratories, Albuquerque, NM, 1987.

[244] R. L. Wong, *Program CICC, flow and heat transfer in cable in conduit conductors - equations and verification*, preprint, Lawrence Livermore National Laboratory, Livermore, CA, 1989.

Chapter 7

The DAE Home Page

7.1 Introduction

Interest in working with DAE's has continued to grow since 1989. DAE models and methods for working with them have become a standard topic at meetings and workshops in several disciplines. Chapters 1–6 developed many of the basic properties of DAE's and numerical methods for their solution. Chapter 7 concerns developments since 1989. There have been some fundamental theoretical developments, particularly in the area of nonlinear DAE's. Additional types of numerical methods have been investigated. There has been a great deal of work concerning various specialized applications, such as mechanics, and algorithms for efficiently solving numerically the DAE's that arise in these applications. A discussion of all of these contributions would take several hundred pages. Instead we will mention what we consider to be some of the major advances that are in keeping with the philosophy of the first six chapters. Numerous references will be made to the recent literature for details on these developments.

In Section 7.2 we shall describe briefly what we consider to be some of the key theoretical developments concerning DAE's since 1989. Section 7.3 summarizes advances in numerical analysis of DAE's, which include extensions of Butcher's tree theory yielding sharper order results for Runge-Kutta methods applied to DAE's, stable forms of symmetric Runge-Kutta methods, methods and theory for long-time integration of constrained Hamiltonian systems, and multistep methods which exploit the DAE structure. The two longest sections of this chapter are those dealing with software. Section 7.4 concerns powerful symbolic and numerical tools that have become available for supporting environments where DAE solution is required. The DAE code DASSL, which won the first Wilkinson Prize for Numerical Software in 1991, was carefully examined in Chapter 6. Several extensions of DASSL which are designed for various applications such as large-scale DAE's have become available and will be described in some detail in Section 7.5.

The use of DAE's in modeling and simulation has continued to increase rapidly. This growth has occurred in many areas and has been fueled both by the desire to solve more complex problems and the increasing use of computer-generated models based on general descriptions of a problem. In particular, in recent years there has been considerable research on constrained mechanical systems which usually lead to index two or index three DAE's. The solution of PDE's by the method of lines, particularly in chemical engineering, has continued to be a driving force behind software research. There also has been an increasing awareness that DAE's can easily be of higher index. This is especially true of prescribed path control problems [289].

There is increasing interest in large complex models involving many different types of equations. The advantages of being able to directly solve these implicit models are encouraging the study of "DAE's" which include systems other than just ODE's and algebraic equations. Integral equations [312], delay systems [257, 291], PDE's [288, 295, 364], discrete systems, and composites of these systems are beginning to be examined. There is also a growing interest in differential equations with algebraic *inequality* constraints, such as those that arise in the field of optimal control [266, 272, 273, 392, 397, 398] or prescribed path control [390]. However, we will continue to focus in this chapter on systems of differential and algebraic equations.

The supplementary bibliography at the end of this chapter contains the references cited in this chapter that are not in the bibliography at the end of Chapter 6. Because of the widespread interest in DAE's, the literature on DAE's is scattered across many disciplines that use differing terminology. To assist the interested reader, we have included a number of additional references in the supplementary bibliography. These additional entries are publications that either we have found of interest or that serve as pointers to the literature where additional information can be found.

7.2 Theoretical Advances

The original theory for linear time-varying and nonlinear DAE's was motivated by the linear time-invariant case, $Ax'+Bx = f$, where A, B are constant matrices; this is discussed in Chapter 2. The understanding of fully-implicit nonlinear systems in recent years has required extensions and modifications of some of these earlier ideas. In particular, there have been developments concerning the index, solvability, linearization, and the structure of linear time-varying systems.

The Index

There were several equivalent characterizations of the index for linear time-invariant DAE's, $Ax'+Bx = f$. The situation is very different in the nonlinear case. For some important special classes of DAE's, it is sufficient to consider

7.2. THEORETICAL ADVANCES

two types of indices, namely the perturbation index and the differentiation index, in order to understand the smoothness properties of solutions and to derive error estimates for numerical methods. However, for more general classes of higher index DAE's, the question of what is the index becomes more complicated, and several definitions are possible. The index, in all its variations, measures in some sense how the solution of the DAE depends on the describing equations, initial data, and forcing functions. A good understanding of the index is essential for the development of numerical integrators that can be used on those DAE's for which BDF or IRK based methods are not appropriate.

In Chapter 2 we defined what is sometimes now called the *differentiation index* ν_d. It is the number of times that all or part of $F(t, y, y') = 0$ must be differentiated to uniquely determine y' in terms of (t, y) for consistent (t, y). The *perturbation index* ν_p is introduced in [136], where it plays a fundamental role in obtaining order results for IRK methods.

Some notation is necessary to define ν_p. Let $\|\cdot\|$ be the usual Euclidean norm on $\mathbf{R}^n$. For integers $m \geq 1$, let $\mathcal{C}^m$ be the space of m times continuously differentiable $\mathbf{R}^n$-valued functions on the finite interval $\mathcal{I}$. Let $\mathcal{C}^0$ be the space of continuous $\mathbf{R}^n$-valued functions on the finite interval $\mathcal{I}$. For $m \geq 0$ we give $\mathcal{C}^m$ the norm $\|g\|_m = \sum_{i=0}^{m} \|g^{(i)}\|_\infty$, where $\|h\|_\infty = \sup_{t \in \mathcal{I}} \|h(t)\|$. For notational convenience let $\|h\|_{-1} = \int_\mathcal{I} \|h(t)\| \, dt$. Let $\|f\|_p^t$ be $\|f\|_p$ on the interval $[0, t]$ for $p \geq -1$.

Definition 7.2.1 *The DAE $F(t, y, y') = 0$ has* perturbation index ν_p *along a solution y on the interval $\mathcal{I} = [0, T]$ if ν_p is the smallest integer such that if*

$$F(t, \widehat{y}, \widehat{y}') = \delta(t) \tag{7.2.1}$$

for sufficiently smooth δ, then there is an estimate

$$\|\widehat{y}(t) - y(t)\| \leq C \left(\|\widehat{y}(0) - y(0)\| + \|\delta\|_{\nu_p - 1}^t \right) \tag{7.2.2}$$

for sufficiently small δ in the $\|\cdot\|_{\nu_p - 1}$ norm. C is a constant that depends on F, the length of the interval, and the solution y.

The perturbation index describes the continuity of the solutions y of the DAE $F(t, y, y') = \delta(t)$ as $\delta \to 0$ in the $\|\cdot\|_{\nu_p - 1}$ norm. For many classes of DAE's ν_p is either the same as ν_d or one higher. In [292] it is shown that ν_p and ν_d can differ by much more than one for fully-implicit nonlinear DAE's. However, it also is shown there that if one defines the *maximum perturbation index*, ν_{MP}, and the *maximum differentiation index*, ν_{MD}, over suitable open sets, then $\nu_{MD} \leq \nu_{MP} \leq \nu_{MD} + 1$. One can think of these maximum indices as giving information about the continuity of the solution y of $F(t, y, y') = \delta$ in terms of δ for small δ and not just as $\delta \to 0$.

Following the notation of (2.5.4), let $\mathbf{F}_j(t, y, y', \ldots, y^{(j)}) = \mathbf{F}_j(t, y, y', w)$ be the *derivative array equations* obtained by differentiating the DAE $j - 1$

times with respect to t. The graph of $(y(t), y'(t), \ldots, y^{(j)}(t))$ is called the extended graph of the function y. An extended neighborhood Ω^e is an open set in (t, y, y', w) space. The *uniform differentiation index* [292] ν_{UD} of the DAE $F(t, y, y') = 0$ on Ω^e is the smallest integer k, if it exists, such that the following four conditions hold on the open set Ω^e:

(A1) Sufficiency smoothness of $\mathbf{F} = \mathbf{F}_{k+1} = 0$.

(A2) Consistency of $\mathbf{F} = 0$ as an algebraic equation.

(A3) $\overline{J}_k = [\ \mathbf{F}_{y'}\ \ \mathbf{F}_w\ ]$ is 1-full and has constant rank.

(A4) $J_k = [\ \mathbf{F}_{y'}\ \ \mathbf{F}_w\ \ \mathbf{F}_y\ ]$ has full row rank independent of (t, y, y', w).

Note that we are not requiring that there be a μ for which (A1)–(A4) hold for $1 \leq k \leq \mu$. Rather we assume that there is a μ such that (A1)–(A4) hold for $k = \mu$. If (A1), (A2), (A4) hold for one value of k, then they will hold for lower values of k. This is not the case with (A3). (A3) holding for $k = \mu$ need not imply that (A3) holds for any $k < \mu$.

Direct computation of ν_{MD} or ν_{MP} can be difficult. However, it is shown in [292] that $\nu_{MD} = \nu_{UD}$. Computation of ν_{UD} is addressed in [293]. Assumptions (A1)–(A4) and ν_{UD} are fundamental to the development of the derivative array-based integrators discussed in the next subsection.

In the remainder of this chapter, unless specified otherwise, the term "the index" will mean the differentiation index. The one exception is when it refers to the derivative array approach, where we will mean ν_{UD}.

Solvability

For fully-implicit higher index DAE's the characterization of the solution manifold is important in the development of numerical integrators. All of the integrators currently under development for fully-implicit DAE's are based in part on one of the approaches for proving that a given DAE is solvable.

The structure of fully-implicit nonlinear DAE's is now much better understood. The theory has been developed in two closely related ways. One approach is to proceed in a manner that is a nonlinear analog of the algorithms for the linear case, which reduce the index by successively differentiating the constraints and substituting from the differential equations to form new constraints. In this "geometric" approach, one constructs a series of manifolds of decreasing dimension using a sequence of constant rank assumptions and tools from differential geometry [370, 371, 381]. The manifold at each stage of the process corresponds to the solution of a set of constraints. The manifold has lower dimension as additional constraints are determined. For low index DAE's, this approach can be used to construct numerical methods which are based on a parameterization of the solution manifold. For some problems

7.2. THEORETICAL ADVANCES

where the solution manifold has a high curvature, the geometric approach can be better than approaches which use a fixed set of local coordinates over time subintervals, such as is done by the coordinate partitioning approach from mechanics [241].

For the linear time-varying case

$$A(t)y' + B(t)y = f(t), \qquad (7.2.3)$$

the conditions (A1)–(A4) are equivalent to (7.2.3) being solvable for every sufficiently smooth $f(t)$ and the solution y being continuous in f in an appropriate function space [58]. This is called *uniform solvability* in [292]. For higher index systems, the assumptions (A1)–(A4) are independent of constant rank assumptions on A in (7.2.3) or $F_{y'}$ in $F(t, y, y') = 0$. The assumptions in [369] permit variable-rank Jacobians and are a special index two case of (A1)–(A4).

An alternative to the geometric approach to nonlinear DAE's is the "derivative array" approach developed in [286, 287, 293]. Solvability can be proven locally if assumptions (A1)–(A4) hold for k and $k+1$ [293]. The assumptions of the derivative array approach are technically independent of the assumptions for the geometric approach since they do not require the intermediate constant rank assumptions. However, this independence is probably not significant in many problems. Of more importance for higher index problems is that the derivative array Jacobian is more easily computed. Integrators for general nonlinear higher index DAE's based on the derivative array approach are under development [297, 357]. Currently, the most promising approach based on the derivative array determines a local set of coordinates for y and then solves the derivative array equations numerically in a least squares sense for y'. This determines numerically an ODE in the local coordinates which is then integrated. Since no constraints are assumed to be known or explicitly computable, this approach is known as *implicit coordinate partitioning*. The approach is designed for moderate-sized dense nonlinear DAE's of "moderate" index. Implicit coordinate partitioning has been successfully used on a variety of fully-implicit nonlinear DAE's of index four, five, and six. Note that one does not actually have to perform any of the differentiation symbolically. All derivatives needed in forming the derivative array and in calculating the Jacobians of the derivative array which are required for the least squares iteration can be found using automatic differentiation software such as ADOL-C [298].

Linearization

The analysis of many numerical methods for DAE's depends on a linearization of the DAE. Linearizations also are important in control analysis and design. For certain special classes of DAE's such as Hessenberg systems, it is known that the linearization inherits that structure and that the linearization can provide stability information. However, the linearization of a general DAE need not even have a solution manifold of the same dimension as the original nonlinear DAE [290]. In [382] it is shown that the linearization about an equilibrium point of an autonomous DAE can provide useful information about the behavior of the original nonlinear DAE near the equilibrium. These results are extended in [290] to show that while a time-invariant linearization of a nonlinear DAE is sometimes incorrect, linear time-varying linearizations along trajectories accurately reflect the nonlinear vector field near the trajectory.

Linear Systems

Linear time-varying DAE's

$$A(t)y'(t) + B(t)y(t) = f(t) \tag{7.2.4}$$

are important for several reasons. They are easier to work with theoretically than fully nonlinear DAE's. Yet the theory for (7.2.4) is closer to the nonlinear case than the linear time-invariant case. Also, the recent result [290] that linear time-varying linearizations can correctly predict some of the behavior of nonlinear DAE's suggests that (7.2.4) will play an even greater role in future analytical and numerical work with nonlinear DAE's.

Theorem 2.4.5 can be improved by setting $G = 0$ in (2.4.5). The proof follows from a consideration of dual operators in [299].

More general types of systems and their solutions have been considered. Distribution and impulsive solutions of $A(t)x'(t) + B(t)x = f(t)$ where f is discontinuous are discussed in [371]. Rectangular systems are carefully considered in [336, 337, 338, 380]. Rectangular systems are especially important in optimization problems where the dynamics have additional degrees of freedom over which some criteria are to be optimized.

Other Developments

Impasse points are places where solutions of a differential equation suddenly end or originate. Impasse points arise in many applications, including circuit and power system models [400]. The study of DAE's, including impasse points and singularities, encompasses all of the issues usually investigated in the framework of bifurcation theory. Mathematical characterizations of impasse points of $A(x)x' = G(x)$ are given in [372] for the index one case. See also [376, 377]. Distributional solutions and discontinuous forcing functions are examined in [374, 375]. A computational algorithm is described in [373] for the case when the underlying ODE has a standard singular point.

7.3 Numerical Analysis Advances

An enormous amount of research on numerical methods for DAE's has taken place since the first writing of this book. Here we point out some significant developments and provide pointers to the relevant literature.

7.3. NUMERICAL ANALYSIS ADVANCES

Butcher's elegant tree theory [283], relating the coefficients of Runge-Kutta methods to their order for ODE's, has been extended to index one, Hessenberg index two, and Hessenberg index three DAE's, yielding sharp order conditions. Many of these results can now be found in the classic text by Hairer and Wanner [323]. Further improvements and extensions to these results are given in [321, 327]. Half-explicit Runge-Kutta methods, which are advantageous for nonstiff DAE's, are developed and analyzed in [271].

A code RADAU5 which is based on the three-stage Radau IIA method is presented in [323]. It is applicable to DAE systems of index one, two, and three of the form
$$By' = f(t, y), \quad y(t_0) = y_0,$$
where B is a constant square matrix which may be singular. For solving higher index systems, the user must be able to identify the higher-index variables. The method is order 5 for index one systems and for the differential variables in Hessenberg index two systems. The code is available by anonymous ftp at ftp.unige.ch (see pub/doc/math/read_me for instructions and further information).

Symmetric[1] Runge-Kutta methods are of interest for boundary value problems (BVP's) in DAE's, where the need to maintain stability for the differential part of the system often necessitates the use of a symmetric method. The formulation of Runge-Kutta methods described in Chapter 4, which was the focus of early research in this area, has problems for symmetric methods applied to fully-implicit index one and higher index Hessenberg DAE's. These problems include instability, order reduction, and persistent oscillations. A class of projected implicit Runge-Kutta methods [256, 346] has been developed which overcomes these problems. Projected collocation methods are particularly advantageous because they retain superconvergence order. These are the methods used in Ascher and Spiteri's code COLDAE [259] for DAE BVP's.

For constrained Hamiltonian systems and constrained conservative mechanical systems, methods preserving certain features of the flow have recently been developed. They encompass partitioned Runge-Kutta methods [328, 330] and composition methods [383] based, for example, on the Rattle-Verlet algorithm [343]. Such methods can retain the symplectic and reversible structures [322, 387] of the flow. They also can achieve a better qualitative behavior than nonpreserving methods. This can be explained by a simple backward analysis argument: the numerical solution given by such methods can be interpreted as the exact solution of a perturbed system possessing the same structure. They also attain a linear growth of the global error with time for integrable and periodic systems, which is desirable for long-time integration, whereas for nonpreserving methods the growth is generally quadratic.

---

[1] If $a_{M-i+1, M-j+1} + a_{ij} = b_{M-j+1} = b_j$, the M-stage IRK is symmetric, e.g., the implicit midpoint or the trapezoid rule.

Extrapolation methods for index one and higher index Hessenberg systems have been developed and analyzed [323]. A code MEXX [345, 346, 347] has been developed by Lubich et al. for solving the equations of motion of constrained multibody systems. MEXX uses extrapolation methods based on half-explicit Euler or half-explicit midpoint formulas. The code exploits the structure of mechanical systems for greater efficiency. Of particular interest in this work is the clear explanation of the "$O(n)$" method. Simeon [388] has developed a library of numerical integration software for constrained mechanical motion called MBSPACK. There has been much work on the formulation of mechanical systems and its effect on numerical stability; see for example [254, 255, 258].

For multistep methods, the thesis of Arévalo [248] considers a class of compound linear multistep methods for semi-explicit index one systems, Hessenberg index two systems, and Hessenberg index three systems. In these methods, the structure of the semi-explicit DAE is exploited by using different discretizations for the differential and algebraic parts of the system. This enables the use of explicit linear multistep methods for semi-explicit DAE systems and also corrects some of the deficiencies of implicit methods such as the $O(1)$ errors in the algebraic variables in index three problems following a change in stepsize or order. A comprehensive theory is presented, and further results are given in [249, 250, 251].

7.4 Software Tools for DAE-Solving Environments

The earlier chapters have shown that many factors influence which integrator should be used on a particular DAE. The move toward complex software environments where computer-generated DAE models are numerically integrated has lead to research on various types of auxiliary software. Among the software tools that the environment should have available are index determination, index reduction, initialization, variable classification (algebraic, index k, etc.), structure determination (e.g., type of problem: Hessenberg, semi-explicit, etc.), singularity determination, and integrator choice. Generally speaking, there are two types of algorithms: graph theoretical (or structure-based) algorithms and analytical algorithms (symbolic or numerical). Symbolic or automatic differentiation is an essential component of many of these software tools. We will begin with a brief description of these technologies as they relate to DAE's.

Symbolic and Automatic Differentiation in DAE's

In the past a scientist could either do hand calculations involving formulas or numerical computations on a computer. There are now more options, and these choices have had an impact on the software tools being developed to work with DAE's. Some of this software will be mentioned in the sections

7.4. SOFTWARE TOOLS FOR DAE-SOLVING ENVIRONMENTS 217

that follow. Symbolic software, such as MAPLE, MACSYMA, REDUCE, DERIVE, and Mathematica, are able to manipulate formulas. Many of these packages also have the options of converting the symbolic results of these manipulations to FORTRAN or C routines and of having this conversion process strive to optimize the resulting code, for example, by introducing new variables for frequently repeated terms, etc. to improve the code efficiency. Sometimes the symbolic software is used to manipulate the model, and then the FORTRAN or C routines are fed to the appropriate integrator.

The computation of derivatives frequently arises when considering DAE's. The derivatives may be taken either during derivation of the equations forming the DAE (as with the Euler-Lagrange equations in mechanics), in the forming of Jacobians (as in implicit integration schemes such as BDF or IRK), or in the derivative array equations. First derivatives may often be estimated using differences. In some cases, derivatives may be found using symbolic software. However, symbolic software often suffers from what is called expression swell. Expressions become very complicated if repeated differentiations are taken. General purpose symbolic codes also tend to be slow and demand large amounts of memory on even moderately large or complex problems. A third option is *automatic differentiation*. Automatic differentiation is like the evaluation of formulas from the symbolic approach in that it gives answers which are correct up to roundoff error. However, it does not have the problems of the symbolic approach with expression swell.

Automatic differentiation is a method for evaluating first and higher partial derivatives of multi-variable functions. Automatic differentiation can handle functions defined by a large variety of procedures or subroutines. An explicit formula is not required. Automatic differentiation also tends to be faster than symbolic approaches and to require substantially less memory. A numeric value is returned for the quantity of interest at a given value of the variables. Automatic differentiation is based on the observation that all functions are evaluated as composites of arithmetic operations and a finite library of nonlinear functions. Through these elementary functions derivatives of arbitrary order can be propagated via the chain rule in the form of truncated Taylor series. This propagation is organized into a highly efficient form using graph theoretic algorithms. The computational savings can be dramatic. The gradient of a scalar function can always be computed using automatic differentiation with no more than 5 times the cost of evaluating just the function itself. This result is *independent of the number of independent variables*. Thus Jacobians can be computed for no more than $5m$ times the work of evaluating the original function where m is the dimension of the vector function. This can alter, for some applications, what is considered to be a practical computation.

Not all automatic differentiation codes have the same capabilities. Two automatic differentiation codes are mentioned in this chapter. ADIFOR works with FORTRAN code. ADOL-C works with C code. ADOL-C also has the capability of taking derivatives with respect to the Taylor coefficients. That is, it can compute $(d^m/dt^m)(Q(y(t))$ and also differentiate $(d^m/dt^m)(Q(y(t))$ with respect to $d^r y/dt^r$.

Graph Theoretical Algorithms for DAE's

The graph theoretical algorithms have received the most study. These algorithms use only the zero pattern of entries and not their actual values. Structural properties were briefly discussed on pages 21 and 22. Let $*$ denote an entry that may be nonzero. Then we would say that the 2×2 matrix $\begin{bmatrix} * & * \\ * & * \end{bmatrix}$ is *structurally nonsingular*. Similarly the pencil $\{ \begin{bmatrix} * & 0 \\ 0 & 0 \end{bmatrix}, \begin{bmatrix} * & * \\ * & * \end{bmatrix} \}$ is *structurally index one*.

Note that the word structure is used in two ways in the DAE literature. For example, a DAE can have the Hessenberg structure of Definition 2.5.3. However, to say that a DAE is structurally Hessenberg does not mean it is actually Hessenberg. Structurally Hessenberg means that the DAE has the zero pattern of a Hessenberg DAE and the products of Jacobians in the definition of a Hessenberg system which are supposed to be nonsingular are structurally nonsingular.

One of the first graph theoretical algorithms for DAE's determined whether the structural index was one [401]. The algorithm in [101] determined if the structural index was one, two, or higher. The algorithm of Pantelides [199] could find consistent initial values, the minimum number of differentiations to determine x_0, x'_0, and the structural index.

The algorithm of [199] has been modified in several ways. By using auxiliary variables one can lower the index without altering the solution set [356]. Also, the algorithm of [199] can be used to determine powers of s in $(Es+A)^{-1}$ [281]. This in turn can be used for scaling in BDF and IRK based methods, index reduction, and index determination. Given a system of equations in some variables and their derivatives, the algorithm of [342] determines a choice of inputs, outputs, and states so that the resulting DAE is structurally index one. Finally, [384] gives a test for structural degeneracy. That is, it will determine whether a DAE is not structurally solvable.

Structural algorithms have proved very useful. However, in general they can only give a lower bound for the index. For example, the DAE

$$\begin{aligned} x' + y' + x + y &= \cos t, \\ x' + y' + x + 2y &= t \end{aligned}$$

has index one but the structural index is zero. Still, structural algorithms often can be useful in those problems where a procedure using differentiation is being employed and the structural index is less than the true index. The algorithms make use of the graph defining the relationship between the variables (used to determine the structural index) to reduce the amount of differentiation needed. The derivative array (2.5.4) is replaced by one where

7.4. SOFTWARE TOOLS FOR DAE-SOLVING ENVIRONMENTS 219

some equations in the DAE are differentiated less than other equations. The size of the problem that must be considered is then reduced, often substantially. Only with those DAE's of structural index zero which have index above zero is no reduction possible in the number of differentiations.

Several large software packages are under development in the chemical engineering community. One such environment is described in [332, 333, 334]. These papers describe structural algorithms for determining the structural index, structural degrees of freedom, minimum needed differentiation, and a list of differentiation/elimination steps needed to reduce the index (based on [101]). They examine when these structural algorithms can fail to give the correct values. In addition, algorithms are developed for both initialization and integration. Another environment from chemical engineering is described in [359]. A MAPLE-based environment in [358] also will determine if there is Hessenberg structure and generate the Jacobians needed for either DASSL or RADAU5.

Analytic Algorithms for DAE's: Symbolic and Numerical

In terms of available software, the graph-based algorithms are probably more highly developed than the analytic-based algorithms. However, in some areas, such as chemical process simulation, the DAE's that result frequently do not have any easily recognized structure [333]. Graph-based algorithms also can underestimate the index in problems dealing with the control of mechanical systems [293]. Some progress has been made on algorithms using various mixtures of symbolic and numeric calculations.

Robust initialization remains a difficult problem in many applications. Initialization is important not only in beginning the integration but in understanding how to handle the solution discontinuities that frequently occur in applications. One of the first numerical approaches to determining initial values for higher index DAE's was described in [162]. It uses numerical differentiation to approximate the derivatives. Rank assumptions similar to those of the derivative array approach are made, as well as assumptions that the linearization of the DAE and the original DAE are appropriately related. This assumes, among other things, that the DAE can be transformed by constant coordinate transformations to one in which the structural index is the same as the index. This restricts somewhat the possible nonlinearities allowed in the DAE. Newton iterations are used in the algorithm. These ideas are further examined in [332, 333]. In those applications the Newton iteration must be carefully damped in order to obtain convergence for solutions which are not near the steady state. The structural assumptions can be relaxed if Jacobians are computed by either symbolic differentiation or automatic differentiation [286, 287]. A continuation approach for initialization is discussed in [349]. However, generating good starting values and correct management of the iteration remains a topic of research [277].

A variety of other information can be computed from the Jacobian of the derivative array. Some of this information, such as the dimension of the solution manifold, can be computed numerically for DAE's which satisfy (A1)-(A4). However, other calculations, such as verifying (A3), are more difficult and at this time can only be actually verified symbolically. However, numerical tests can be done, which while not actually verifying (A3) usually suffice to determine in practice whether (A3) holds. Among the calculations that can be done from the derivative array are showing that the system is solvable, computing the index (namely, ν_{UD}), and determining the dimension of the solution manifold [293]. A robotic arm path control example is given in [293] where structural calculations underestimate the index by two but the numerical approach developed there correctly computes all the needed information.

In some nonlinear DAE's there is, in fact, an advantageous structure to the DAE but this structure is obscured by the coordinates used to describe the system. If the coordinates used are related to the "nice coordinates" by a (possibly unknown) constant coordinate change, then numerical algorithms for determining the Kronecker structure of a matrix pencil can be useful [305, 306].

7.5 The DASSL Family

Several new codes which significantly extend the capabilities of DASSL have been developed during the past few years. These include DASPK for large-scale DAE systems, parallel versions of DASPK, and DASSLSO and DASPKSO for sensitivity analysis of DAE systems. In this section we will describe the new software. We will also describe recent improvements to the algorithms in DASPK for finding consistent initial conditions. We end this section by giving details on the availability of DASSL and its extensions.

DASPK for Large-Scale DAE Systems

DASPK was developed by Brown, Hindmarsh, and Petzold [276] for the solution of large-scale systems of DAE's. It is particularly effective in the method-of-lines solution of time-dependent PDE systems in two or three dimensions. In contrast to DASSL, which is limited in its linear algebra to dense and banded systems, DASPK makes use of the preconditioned GMRES [386] iterative method for solving the linear system at each Newton iteration. Here we describe some of the basic features of the DASPK algorithm. Further details on the structure and use of DASPK and on a class of preconditioners for DAE systems for reaction–diffusion type problems can be found in [276].

DASPK combines the time-stepping methods of DASSL with the preconditioned iterative method GMRES for solving the linear system at each Newton

7.5. THE DASSL FAMILY

iteration. The integration methods and strategies for time-stepping are virtually identical to those in DASSL, as described in Chapter 5. DASPK also has an option to use direct methods (as in DASSL) which can be useful for debugging on small prototype systems.

Differences between DASSL and DASPK occur in the algorithms for solving the nonlinear and linear systems and the convergence test. Recall that following discretization by the BDF methods, a nonlinear equation

$$F(t, y, \alpha y + \beta) = 0 \qquad (7.5.5)$$

must be solved at each time step, where $\alpha = \alpha_0/h_n$ is a constant which changes whenever the stepsize or order changes, β is a vector which depends on the solution at past times, and t, y, α, β are evaluated at t_n. To simplify the discussion, we will sometimes refer to the above function as $F(y)$. Both DASPK and DASSL solve this equation by a modified version of Newton's method,

$$y^{(m+1)} = y^{(m)} - c \left(\alpha \frac{\partial F}{\partial y'} + \frac{\partial F}{\partial y} \right)^{-1} F(t, y^{(m)}, \alpha y^{(m)} + \beta). \qquad (7.5.6)$$

In the case of DASPK with direct methods specified, which is virtually identical to using DASSL, the iteration matrix

$$A = \alpha \frac{\partial F}{\partial y'} + \frac{\partial F}{\partial y} \qquad (7.5.7)$$

is computed and factored and is then used for as many time steps as possible. By contrast, in the iterative methods option, a preconditioner matrix P, which is an approximation to A that leads to a cheap linear system solution, is computed and factored and used for as many time steps as possible. It is often possible to use a preconditioner over more steps than it would be possible to keep an iteration matrix in the direct option, because the iterative methods do the rest of the work in solving the system. One of the powerful features of the iterative approach is that it does not need to compute and store the iteration matrix A explicitly.[2] This is because the GMRES method, as we will see below, never actually needs this matrix explicitly. Instead, it requires only the action of A times a vector v. In DASPK, this matrix–vector product is approximated via a difference quotient on the function F in (7.5.5)

$$Av = F'(y)v \approx \frac{F(t, y + \sigma v, \alpha(y + \sigma v) + \beta) - F(t, y, \alpha y + \beta)}{\sigma}. \qquad (7.5.8)$$

The GMRES algorithm requires products Av in which v is a vector of unit length (the norm is a weighted norm based on the user-defined error tolerances

---

[2]Depending on the preconditioner, it may need to compute and store a preconditioner matrix explicitly. We hope that this matrix is much cheaper to generate and to store than the actual iteration matrix.

as described in Chapter 5) and y is the current iterate. In DASPK, σ is taken to be 1, as we will explain later. We note that, because y is current in (7.5.8), this amounts to taking a full Newton iteration in the iterative option of DASPK (rather than modified Newton, as in DASSL and the direct option of DASPK). In fact, for some highly nonlinear problems we have seen the iterative option in DASPK outperform the direct option in terms of time steps, corrector failures, etc., apparently for this reason.

Choice of the constant c, which is used to speed up the iteration if the current Jacobian approximation is based on a value of α from some previous time step in the direct method, is described in Chapter 5. For the iterative method, while the preconditioner may be based on an old α, the linear system to be solved is based on the current α, so in this case $c \equiv 1$.

The rate of convergence ρ of (7.5.6) is estimated whenever two or more iterations have been taken by

$$\rho = \left(\frac{||y^{(m+1)} - y^{(m)}||}{||y^{(1)} - y^{(0)}||} \right)^{1/m} \tag{7.5.9}$$

(The norms are scaled norms which depend on the error tolerances specified by the user.) The Newton iteration is taken to have converged when

$$\frac{\rho}{1-\rho} ||y^{(m+1)} - y^{(m)}|| < 0.33. \tag{7.5.10}$$

If $\rho > 0.9$ or $m > 4$, and the iteration has not yet converged, then the stepsize is reduced, and/or an iteration matrix based on current approximations to y, y', and α is formed, and the step is attempted again. If the difference between the predictor and the first correction is very small (for the direct solver, this is relative to roundoff error in y; for the iterative solver, it is relative to the accuracy requested in solving the linear system), the iteration is taken to have converged (because the initial correction is so small that it is impossible to get a good rate estimate).

For the iterative methods, convergence tests such as (7.5.10) need to be carefully justified, because the Newton iterates are not computed exactly but instead with a relatively large error, which is due to solving the linear system inexactly. The test (7.5.10) can be justified, at least to some extent, by considering the Newton/GMRES method in the framework of the theory of inexact Newton methods [304]. In this framework, the Newton iteration for $F(y) = 0$, including errors $r^{(m)}$ due to solving the linear system inexactly, is written as

$$F'(y^{(m)}) \delta y^{(m)} = -F(y^{(m)}) + r^{(m)}, \tag{7.5.11a}$$
$$y^{(m+1)} = y^{(m)} + \delta y^{(m)}. \tag{7.5.11b}$$

Theorem 2.2 in Brown and Hindmarsh [274] justifies a convergence test based on rate of convergence for the inexact Newton iteration, provided the residuals

7.5. THE DASSL FAMILY

$r^{(m)}$ satisfy $||r^{(m)}|| \leq \eta ||F(y^{(m)})||$, for $\eta < 1$. However, as in the case of ODE's, we prefer to terminate the linear iteration based on a condition like $||r^{(m)}|| \leq \delta$. In this case, it can be argued heuristically as in [274] that the test is still justified, provided that $\delta << \epsilon/\lambda$, where ϵ is the tolerance for the final computed Newton iterate, $||y^{(m+1)} - y^*|| \leq \epsilon$, and $\lambda = ||(F')^{-1}(y^*)||$. Now, this is almost what we need, except that for most DAE's (and stiff ODE's), λ is likely to be quite large, which would seem to mandate a quite conservative test for the linear iteration. To see how the theory can be used to justify a less conservative test, multiply $F(y) = 0$, and hence (7.5.11a), by P^{-1}, where P is the preconditioner matrix described below. This changes nothing in terms of the Newton iterates or the final solution for y^*. Now, *assuming that the preconditioner is a good approximation to F'*, in the sense that $||P^{-1}F'|| \approx O(1)$, the theory and heuristic arguments [274], applied to $P^{-1}F$ instead of F, justify a rate of convergence-based termination criteria for the Newton iteration, provided the *preconditioned* residuals for the linear iteration satisfy $||P^{-1}r^{(m)}|| \leq \delta$, where $\delta << \epsilon$. In DASPK, we take $\epsilon = .33$, and take $\delta = .05\epsilon$ as the default tolerance for solving the linear iteration (the constant δ/ϵ can be adjusted optionally by the user). The norms used are weighted norms based on the user-defined error tolerances. We note that another desirable property of these tests is that they are invariant under scalings of the DAE (i.e., multiplying F on the left by some arbitrary matrix), provided the preconditioner also has been scaled accordingly.

In contrast to the ODE solvers LSODPK [275] and VODPK [285], which use preconditioned Krylov methods with left and/or right preconditioning, the DASPK solver allows only left preconditioning. The reason has to do with a basic difference between ODE's and DAE's. For a DAE system defined by $F(t, y, y') = 0$, the components of the vector F need not have any relation to those of y. For example, the two vectors need not have the same physical units in corresponding components. If a left preconditioner P_1 and a right preconditioner P_2 are allowed in the solution of the linear system $Ax = b$, where $A = F'$ and b is a value of $-F$, then the Krylov method in effect deals with the matrix $P_1^{-1}AP_2^{-1}$ and with residual vectors $P_1^{-1}r$ ($r = b - Ax$), and performs a convergence test on weighted norms of those vectors. But consistent choices of P_1 and P_2 are possible, with $P_1P_2 \approx A$, for which $P_1^{-1}r$ does not have the same units as y. Then norms of the preconditioned residuals $P_1^{-1}r$ are meaningless as a measure of the quality of the current approximate solution vector x. In contrast, if $P_2 = I$ and P_1 is a consistent approximation to A, then $P_1^{-1}r$ has the same units as y in each component, and the convergence test in the Krylov algorithm, with the same weighted norm as used in the local error test, is completely consistent. Moreover, that convergence test is invariant under a change of scale in either the function F or the vector y (provided the absolute tolerances are rescaled consistently if y is). This consistency and scale invariance are not possible with preconditioning either on the right only or on both sides.

In DASPK, the iterative option *requires* the user to provide a preconditioner P. This is in part because the Newton iteration test, and hence ultimately the code reliability, is not justified without a halfway-reasonable preconditioner. It also is because *any nontrivial DAE needs a preconditioner*. Even a "nonstiff" DAE needs a preconditioner, to approximate the Jacobian of the constraint matrix. Also, recall from Section 5.4.2 that the iteration matrix for any nontrivial DAE becomes more and more ill conditioned as the stepsize is reduced. Therefore, the preconditioner may need to rescale the system as discussed in Section 5.4.2. In the case that the DAE is really an ODE, with $F(t, y, y') = y' - f(t, y)$, the preconditioner could be taken as $P \approx \alpha * I$, although even for ODE's it is often better to provide a nontrivial preconditioner [275].

Solving (7.5.6) requires the solution of a linear system

$$Ax = b \qquad (7.5.12)$$

at each Newton iteration, where A is the $N \times N$ iteration matrix in (7.5.7), $x = y^{(m+1)} - y^{(m)}$ is an N-vector, and $b = -cF(t, y^{(m)}, \alpha y^{(m)}) + \beta)$ is an N-vector. Solution in the direct option is described in Section 5.2.2.

In the case of iterative methods, the linear system (7.5.12) is solved by the preconditioned GMRES iterative method [386]. Depending on the options chosen, the method may be either the complete or the incomplete GMRES method. It may be restarted.

GMRES is one of a class of *Krylov subspace projection methods* [385]. The basic idea of these methods is as follows. If x_0 is an initial guess for the solution, then letting $x = x_0 + z$, we get the equivalent system $Az = r_0$, where $r_0 = b - Ax_0$ is the initial residual. We choose $z = z_l$ in the *Krylov subspace* $K_l = \text{span}\{r_0, Ar_0, \ldots, A^{l-1}r_0\}$. For the GMRES algorithm, z_l, and hence $x_l = x_0 + z_l$, is specified uniquely by the condition

$$\|b - Ax_l\|_2 = \min_{x \in x_0 + K_l} \|b - Ax\|_2 \quad (= \min_{z \in K_l} \|r_0 - Az\|_2).$$

Here, $\|\cdot\|_2$ denotes the Euclidean norm and $<.,.>$ the corresponding inner product.

GMRES uses the *Arnoldi process* [253] to construct an orthonormal basis of the Krylov subspace K_l. This results in an $N \times l$ matrix $V_l = [v_1, \ldots, v_l]$ and an $l \times l$ upper Hessenberg matrix H_l such that

$$H_l = V_l^T A V_l \text{ and } V_l^T V_l = I_l \quad (= l \times l \text{ identity matrix}).$$

If the vectors $r_0, Ar_0, \ldots, A^l r_0$ are linearly independent, so that the dimension of K_{l+1} is $l+1$, then the matrices $V_{l+1} = [v_1, \ldots, v_{l+1}]$ and $\bar{H}_l \in \mathbf{R}^{(l+1) \times l}$ defined by

$$\bar{H}_l = \begin{bmatrix} H_l \\ r^T \end{bmatrix}, \text{ where } r = [0, \ldots, 0, h_{l+1,l}]^T \in \mathbf{R}^l,$$

7.5. THE DASSL FAMILY

satisfy
$$AV_l = V_{l+1}\bar{H}_l.$$

Furthermore, letting $z = V_l y$, we find that $\|r_0 - Az\|_2 = \|\beta e_1 - \bar{H}_l y\|_2$, where $\beta = \|r_0\|_2$ and e_1 is the first standard unit vector in $\mathbf{R}^{l+1}$. The vector $y = y_l$ minimizing this residual is computed by performing a QR factorization of $\bar{H}_l$ using Givens rotations. Then the GMRES solution is $x_l = x_0 + V_l y_l$. As noted by Saad and Schultz [386], this QR factorization can be done progressively as each column appears, and one can compute the residual norm $\|b - Ax_l\|_2$ without computing x_l at each step. If the $\sin\theta$ elements of the Givens rotations are denoted s_j $(j = 1, \ldots, l)$, then one obtains

$$\|b - Ax_l\|_2 = \beta |s_1 \cdots s_l|. \tag{7.5.13}$$

The use of (7.5.13) leads to the following algorithm, in which $l_{\max}$ and δ are given parameters.

Algorithm GMRES

1. Compute $r_0 = b - Ax_0$ and set $v_1 = r_0/\|r_0\|_2$.

2. For $l = 1, \ldots, l_{\max}$ do:

    (a) Form Av_l and orthogonalize it against $v_1, \ldots, v_l$ via
    $$w_{l+1} = Av_l - \sum_{i=1}^{l} h_{il} v_i, \quad h_{il} = <Av_l, v_i>$$
    $$h_{l+1,l} = \|w_{l+1}\|_2$$
    $$v_{l+1} = w_{l+1}/h_{l+1,l}.$$

    (b) Update the QR factorization of $\bar{H}_l$.

    (c) Use (7.5.13) to compute $\rho_l = \|r_0\|_2 \cdot |s_1 \cdots s_l| = \|b - Ax_l\|_2$.

    (d) If $\rho_l \leq \delta$, go to Step 3. Otherwise, go to (a).

3. Compute $x_l = x_0 + V_l y_l$, and stop.

In the above algorithm, if the test on ρ_l fails, and if $l = l_{\max}$ iterations have been performed, then one has the option of either accepting the final approximation x_l or setting $x_0 = x_l$ and then going back to Step 1 of the algorithm. This last procedure has the effect of "restarting" the algorithm, and DASPK does such restarts when necessary to achieve convergence.

As l gets large, much work is required to make v_{l+1} orthogonal to all the previous vectors $v_1, \ldots, v_l$. One can propose an incomplete version of GMRES (denoted by IGMRES), which differs from Algorithm GMRES only in that the sum in Step 2(a) begins at $i = i_0$ instead of at $i = 1$, where $i_0 = \max(1, l - p + 1)$. Details are given in [275].

Saad and Schultz [386] have given a convergence analysis of Algorithm GMRES which shows that the GMRES iterates converge to the true solution of (7.5.12) in at most N iterations. We also note that Algorithm GMRES may have breakdowns. If $w_{l+1} = 0$ in the Arnoldi process, then Saad and Schultz [386] have shown x_l is the exact solution of (7.5.12). This is referred to as a happy breakdown. When $w_{l+1} \neq 0$, the matrix $\bar{H}_l$ has full column rank, and so the least squares problem for y_l can always be solved via the above QR factorization. (However, in some cases the approximation x_l may not be of much use. An example is given in [275] illustrating how GMRES can have a dramatic failure.)

Realistic DAE problems require the inclusion of scale factors, so that all vector norms become weighted norms in the problem variables. However, even the scaled iterative methods seem to be competitive only for a fairly narrow class of problems, namely ODE's characterized mainly by tight clustering in the spectrum of the system Jacobian. Thus for robustness, it is essential to enhance the methods further. As in other contexts involving linear systems, preconditioning of the linear iteration is a natural choice. In what follows, the use of scaling and preconditioning in DASPK is described.

The user of DASPK must provide parameters that define error tolerances to be imposed on the computed solution. These are relative and absolute tolerances RTOL and ATOL such that the combination

$$w_i = RTOL_i |y_{i,n-1}| + ATOL_i$$

is applied as a scale factor for component y_i during the time step from t_{n-1} to t_n. Specifically, a weighted root-mean-square norm

$$\|x\|_{WRMS} = \left[N^{-1} \sum_1^N \left(\frac{x_i}{w_i} \right)^2 \right]^{1/2}$$

is used on all error-like vectors. Thus if we define a diagonal matrix

$$D = \sqrt{N} \text{diag}(w_1, \ldots, w_N),$$

we can relate this to an ℓ_2 norm:

$$\|x\|_{WRMS} = \|D^{-1}x\|_2.$$

Because D contains the tolerances, the local error test on a vector e of estimated local errors is simply $\|e\|_{WRMS} \leq 1$.

The linear system in (7.5.12) can be restated in scaled form in terms of $D^{-1}x = \tilde{x}$ and $D^{-1}b = \tilde{b}$. Likewise, the nonlinear system $F(y) = 0$ can be restated in a scaled form $\tilde{F}(\tilde{y}) = 0$.

We note that, while DASSL allows the user to replace the norm subroutine, DASPK does not allow this. This is because a scaled ℓ_2 norm is needed in

7.5. THE DASSL FAMILY

the implementation of the GMRES algorithm. However, DASPK does allow a user-replaceable subroutine to define the weights in the norm. (The default is to set the weights according to the tolerances RTOL and ATOL via EWT(I) = RTOL(I)*ABS(Y(I)) + ATOL(I)). We recommend that changing the weights be attempted only after careful thought and consideration.

When a basic iteration fails to show acceptable convergence on a given problem, preconditioning is often beneficial, especially when the cause of the slow convergence can be identified with one or more parts of the problem which are (individually) easier to deal with than the whole problem. Generally, preconditioning in an iterative method for solving $Ax = b$ means applying the method instead to the equivalent system

$$(P^{-1}A)(x) = (P^{-1}b), \text{ or } \bar{A}x = \bar{b}, \tag{7.5.14}$$

where P is chosen in advance. The preconditioned problem is easier to solve than the original problem provided that (1) linear systems $Px = c$ can be solved economically, and (2) P is in some way close to A. Condition (1) is essential because carrying out the method on $\bar{A}x = \bar{b}$ clearly requires evaluating vectors of the form $P^{-1}c$, at the beginning of the iteration, during each iteration, and at the end. Condition (2) is less well defined, but means simply that the convergence of the method for $\bar{A}x = \bar{b}$ should be much better than for $Ax = b$, because $\bar{A}$ is somehow close to the identity matrix (for which convergence is immediate).

It is essential that the scaling of the linear system (discussed above) be retained in the preconditioned methods. Since the scaling matrix D is based on the tolerance inputs to the integrator, D^{-1} can be thought of as removing the physical units from the components of x so that the components of $D^{-1}x$ can be considered dimensionless and mutually comparable. On the other hand, the matrix $A = \alpha \partial F / \partial y' + \partial F / \partial y$ is *not* similarly scaled, and so, because P is based on approximating A, the matrix

$$\bar{A} = P^{-1}A$$

also is not similarly (dimensionally) scaled. More precisely, it is easy to show that if the (i, j) elements of P each have the same physical dimension as that of A, i.e., the dimension of F_i/y_j, then the (i, j) element of $\bar{A}$ has the dimension of y_i/y_j. Similarly, for the vectors x and $\bar{b}$, the ith component of each has the same physical dimension as that of y_i. It follows that the diagonal scaling D^{-1} should be applied to x and $\bar{b}$ in the same way that it was applied to x and b without preconditioning. Thus we change the system (7.5.14) again to the equivalent scaled preconditioned system

$$(D^{-1}\bar{A}D)(D^{-1}x) = (D^{-1}\bar{b}), \text{ or } \tilde{A}\tilde{x} = \tilde{b}. \tag{7.5.15}$$

Combining the two transformations, we have

$$\tilde{A} = D^{-1}P^{-1}AD, \quad \tilde{x} = D^{-1}x, \quad \tilde{b} = D^{-1}P^{-1}b. \tag{7.5.16}$$

In implementing the GMRES method in DASPK, many of the algorithmic issues that arise are the same as for the ODE case. We have carried over the treatment of these matters from LSODPK [275]. Below is a summary of those details.

- DASPK takes $x_0 = 0$, having no choice readily available that is clearly better.

- The scaling is incorporated in an explicit sense, storing vectors $\tilde{v}_i$ that arise in the method as it stands, rather than unscaled vectors $D\tilde{v}_i = v_i$.

- DASPK uses a difference quotient representation
$$Av \approx \frac{F(t, y + \sigma v, \alpha(y + \sigma v) + \beta) - F(t, y, \alpha y + \beta)}{\sigma}.$$

- DASPK takes $\sigma = 1$ because $\|v\|_{WRMS} = 1$. Thus the perturbation vector v can be regarded as a small correction to y, since its WRMS norm $(= 1)$ is a value that is accepted for local errors in y in the local error test.

- The modified Gram-Schmidt procedure is used for orthogonalizing basis vectors.

- DASPK handles breakdowns in the same manner as in the basic algorithms.

- The convergence test constant δ used as a bound on the residuals $\|b - Ax_l\|_{WRMS}$ is taken to be $\delta = .05\epsilon$, where $\epsilon = .33$ is the tolerance on the nonlinear iteration in (7.5.10).

- DASPK uses the complete GMRES method (i.e., $p = l_{\max}$), but the user may change p (called KMP in the code) to be less than $l_{\max}$ to obtain the incomplete GMRES method.

We can now state the algorithm used in DASPK for scaled preconditioned versions of the GMRES method. This is given for arbitrary x_0, for the sake of generality, and is denoted SPIGMR.

Scaled Preconditioned Incomplete GMRES (SPIGMR)

1. (a) $r_0 = b - Ax_0$; stop if $\|r_0\|_{WRMS} < \delta$.
   (b) $\tilde{r}_0 = D^{-1}P^{-1}r_0$, compute $\|\tilde{r}_0\|_2 = \|P^{-1}r_0\|_{WRMS}$, $\tilde{v}_1 = \tilde{r}_0 / \|\tilde{r}_0\|_2$.

2. For $l = 1, 2, \cdots, l_{\max}$, do:
   (a) Compute $\tilde{A}\tilde{v}_l = D^{-1}P^{-1}AD\tilde{v}_l$.

7.5. THE DASSL FAMILY

    (b) $\tilde{w}_{l+1} = \tilde{A}\tilde{v}_l - \sum_{i=i_0}^{l} \tilde{h}_{il}\tilde{v}_i$, where $i_0 = \max(1, l-p+1)$,

        $\tilde{h}_{il} = <\tilde{A}\tilde{v}_l, \tilde{v}_i>$.

    (c) $\tilde{h}_{l+1,l} = \|\tilde{w}_{l+1}\|_2$, $\tilde{v}_{l+1} = \tilde{w}_{l+1}/\tilde{h}_{l+1,l}$.

    (d) Update QR factorization of $\tilde{H}_l = (\tilde{h}_{ij})$ (an $(l+1) \times l$ matrix).

    (e) Compute residual ρ_l indirectly (by (7.5.13) in the complete case).

    (f) If $\rho_l < \delta$, go to Step 3; otherwise, go to (a).

3. Compute $\|\tilde{r}_0\|_2 Q_l^T e_1 = (\bar{g}_l, g)^T$, $\tilde{z} = \tilde{V}_l \bar{R}_l^{-1} \bar{g}_l$, $x_l = x_0 + D\tilde{z}$.

Two parallel versions of DASPK have been developed: DASPKF90, a Fortran 90 data parallel implementation, and DASPKMP, a message-passing implementation written in Fortran 77 with extended BLAS. The parallel versions have been implemented for the Thinking Machines Corporation's CM-5, a massively parallel multiprocessor. Descriptions of the codes and results are given in [348].

DASSLSO and DASPKSO for DAE Sensitivity Analysis

Sensitivity analysis of a DAE model may yield information useful for parameter estimation, optimization, process sensitivity, model simplification, and experimental design. A sensitivity analysis capability has been developed by Maly and Petzold [350] and implemented into DASSL and DASPK, yielding two new codes, DASSLSO and DASPKSO. Here we describe and motivate the algorithms used by the sensitivity option; further details about the codes and some numerical results can be found in [350].

To illustrate the basic approach, consider the general DAE system with parameters

$$F(t, y, y', p) = 0, \quad y(0) = y_0,$$

where $y \in \mathbf{R}^{n_y}$, $p \in \mathbf{R}^{n_p}$. Here, n_y is the number of time-dependent variables y, as well as the dimension of the DAE system, and n_p is the number of (time-independent) parameters in the original DAE system. Sensitivity analysis entails finding the derivative of the solution y with respect to each parameter. This produces an additional $n_s = n_p \cdot n_y$ sensitivity equations which, together with the original system, yield

$$F(t, y, y', p) = 0,$$
$$\frac{\partial F}{\partial y} s_i + \frac{\partial F}{\partial y'} s_i' + \frac{\partial F}{\partial p_i} = 0, \quad i = 1, \ldots, n_p, \quad (7.5.17)$$

where $s_i = dy/dp_i$. Defining

$$Y = \begin{bmatrix} y \\ s_1 \\ \vdots \\ s_{n_p} \end{bmatrix}, \quad F = \begin{bmatrix} F(t,y,p) \\ \dfrac{\partial F}{\partial y}s_1 + \dfrac{\partial F}{\partial y'}s_1' + \dfrac{\partial F}{\partial p_1} \\ \vdots \\ \dfrac{\partial F}{\partial y}s_{n_p} + \dfrac{\partial F}{\partial y'}s_{n_p}' + \dfrac{\partial F}{\partial p_{n_p}} \end{bmatrix}$$

the combined system can be rewritten as

$$\mathbf{F}(t, Y, Y', p) = 0, \quad Y(0) = \begin{bmatrix} y_0 \\ \dfrac{dy_0}{dp_1} \\ \vdots \\ \dfrac{dy_0}{dp_{n_p}} \end{bmatrix}$$

We note that the initial conditions for this DAE system must be chosen to be consistent.

Approximating the solution to the combined system by a numerical method, for example, the implicit Euler method with stepsize h, yields the nonlinear system

$$G(Y_{n+1}) = \mathbf{F}\left(t_{n+1}, Y_{n+1}, \frac{Y_{n+1} - Y_n}{h}, p\right) = 0. \tag{7.5.18}$$

Newton's method for the nonlinear system produces the iteration

$$Y_{n+1}^{(k+1)} = Y_{n+1}^{(k)} - \mathbf{J}^{-1} G(Y_{n+1}^{(k)}),$$

where

$$\mathbf{J} = \begin{bmatrix} J & & & \\ J_1 & J & & \\ J_2 & 0 & \ddots & \\ \vdots & 0 & 0 & J \end{bmatrix} \tag{7.5.19}$$

and

$$J = \frac{1}{h}\frac{\partial F}{\partial y'} + \frac{\partial F}{\partial y}, \quad J_i = \frac{\partial J}{\partial y}s_i + \frac{\partial J}{\partial p_i}.$$

A number of codes for ODE's and DAE's solve the sensitivity system (7.5.17), or its special case for ODE's, directly (see [66] and [379]). If the partial derivative matrices are not available analytically, they are approximated by finite differences. The nonlinear system is usually solved by a staggered scheme, where the first block is solved for the state variables y via Newton's method, and then the block-diagonal linear system for the sensitivities s is solved at each time step.

7.5. THE DASSL FAMILY

Although the direct solution of (7.5.17) is successful for many problems, there are a number of properties of this approach which are not advantageous in the context of DASSL/DASPK. For efficiency, DASSL was designed to use its approximation to the system Jacobian over as many time steps as possible. However, sensitivity implementations using the above staggered approach must reevaluate this Jacobian at every step in order to ensure an accurate approximation to the sensitivity equations. If this matrix has been approximated via finite differences, which is most often the case, large errors may be introduced into the sensitivities. In addition, the staggered scheme for solving the nonlinear system is not advantageous for parallel computation [348]. To eliminate these problems, we focus on approximating the sensitivity system (7.5.17) directly, rather than via the matrices $\partial F/\partial y$, $\partial F/\partial y'$, and $\partial F/\partial p$. In the simplest case, the user can specify directly the residual of the sensitivity system at the same time as the residual of the original system. Eventually, we intend to incorporate the automatic differentiation software ADIFOR [268] for this purpose. Alternatively, we can approximate the sensitivity equations via a directional derivative finite difference approximation. As an example using one-sided differences, the ith sensitivity equation may be approximated as

$$\frac{F(t, y + \delta_i s_i, y' + \delta_i s'_i, p + \delta_i e_i) - F(t, y, y', p)}{\delta_i} = 0, \quad (7.5.20)$$

where δ_i is a small scalar quantity, and e_i is the ith unit vector. Proper selection of the scalar δ_i is crucial to maintaining acceptable round-off and truncation error levels, and will be discussed later. By Taylor's theorem, it is easily seen that (7.5.20) approximates the ith sensitivity in (7.5.17) with an error of order $O(\delta_i)$. We also can approximate the sensitivity system via a second-order central difference with an error of order $O(\delta_i^2)$. Using either of the latter two strategies, approximations to the sensitivity equations are generated at the same time as the residual of the original system, via n_p additional calls to the user function routine. The resulting system is discretized by a numerical method, yielding an iteration matrix of the form (7.5.19).

In general, for a Newton or Newton-Krylov iteration, one should be able to approximate the iteration matrix $\mathbf{J}$ by its block diagonal part provided that the error matrix for the Newton/modified Newton steps is nilpotent. To illustrate this idea, consider the problem formulation (7.5.18)

$$G(Y) = 0$$

and apply a Newton step

$$Y^{(k+1)} = Y^{(k)} - \hat{\mathbf{J}}^{-1} G(Y^{(k)}), \quad (7.5.21)$$

where the Newton matrix $\mathbf{J}$ has been approximated by its block-diagonal part, $\hat{\mathbf{J}}$. The true solution Y^* satisfies

$$Y^* = Y^* - \hat{\mathbf{J}}^{-1} G(Y^*). \quad (7.5.22)$$

Upon subtracting (7.5.22) from (7.5.21) and defining $e^k = Y^{(k+1)} - Y^*$, the iteration errors satisfy

$$Y^{(k+1)} - Y^* = e^{k+1} = e^k - \hat{\mathbf{J}}^{-1}\mathbf{J}e^k = (I - \hat{\mathbf{J}}^{-1}\mathbf{J})e^k$$

to $O(\|e^k\|^2)$ accuracy. The error matrix has the form

$$I - \hat{\mathbf{J}}^{-1}\mathbf{J} = -\begin{bmatrix} 0 & & & & \\ J^{-1}J_1 & 0 & & & \\ J^{-1}J_2 & 0 & \ddots & & \\ \vdots & \vdots & & \ddots & \\ J^{-1}J_{n_p} & 0 & \cdots & \cdots & 0 \end{bmatrix}.$$

It is shown in [350] that because this matrix is nilpotent, the Newton iteration achieves 2-step quadratic convergence for nonlinear problems. Rapid convergence of the Krylov iteration in DASPK using the block-diagonal preconditioner for the sensitivity system also has been observed [350].

The increment selection δ_i is critical to the success of the finite difference approximation of the sensitivity equations. The adaptive strategy used in DASSLSO is described and motivated below.

In the absence of scaling problems, one might choose the increment based on the size of the parameter. For example,

$$\delta_i = \sqrt{u}|p_i|, \qquad (7.5.23)$$

where u is the unit roundoff error, perturbs half the digits of p_i. This type of strategy usually works provided p_i is not near zero. An important consideration for any increment selection strategy is that it should scale with p_i. That is, if the user were to change the units of the parameter and solve instead for $\hat{p}_i = cp_i$, the increment should scale as well. Also we will assume in any such scaling that the error tolerances corresponding to $s_i = dy/dp_i$ have been scaled appropriately. Further indication of the scale of the problem is the relative size of s_i to y. Recall that the directional difference increments y to obtain $y + \delta_i s_i$. Using a rule of thumb which suggests perturbing half the digits of y [316], this would mean roughly that

$$\delta_i \|s_i\|_2 = \sqrt{u}\|y\|_2,$$

and hence

$$\delta_i = \sqrt{u}\frac{\|y\|_2}{\|s_i\|_2}. \qquad (7.5.24)$$

Note also that units of s are the same as units of y divided by units of p. This yields a δ_i with units which are compatible with (7.5.23) and may in some instances yield additional information which can be used by the perturbation

7.5. THE DASSL FAMILY

selection strategy. However, (7.5.24) will fail if $\|s_i\|_2 = 0$, as it often is at the initial time. There also is a potential scaling problem for y, i.e., if $\|y\|_2$ is near zero (for an illustration of scaling problems see the Batch-Reactor numerical results in [350]). Assuming that the user has scaled the error tolerances appropriately, we modify (7.5.24) to prevent such failures and scaling difficulties by using

$$\delta_i = \sqrt{u}\|v_i\|_2, \quad i = 1, \ldots, n_p, \quad (7.5.25)$$

where

$$(v_i)_j = \frac{RTOL_j \cdot |y_j| + ATOL_j}{RTOL_{in_y+j} \cdot |s_{ij}| + ATOL_{in_y+j}} = \frac{WT_j}{WT_{in_y+j}}, \quad j = 1, \ldots, n_y. \quad (7.5.26)$$

Combining (7.5.23) and (7.5.25), the strategy used in DASSLSO is given by

$$\delta_i = \Delta \cdot \max(|p_i|, \|v_i\|_2). \quad (7.5.27)$$

In our experience, it is possible that for some very nasty problems $\Delta = \sqrt{u}$ may be too small. The user is given the option of changing this value. It should be noted that for many well-scaled problems, the perturbation $\Delta = \sqrt{RTOL}$ may be appropriate. This implies that we are perturbing roughly half of the (locally) accurate digits of the numerical solution, hence the error in the second-order difference to the sensitivities is $O(\Delta^2) = O(RTOL)$.

DASSLSO has been designed with the user of DASSL in mind. The parameter list is identical to that of DASSL and all information necessary for sensitivity analysis computation is passed into DASSLSO via the INFO, RPAR, IPAR, RWORK, and IWORK vectors in a manner similar to that of DASSL. When a sensitivity computation has been specified (by setting INFO(16) = n_p), DASSLSO works with the $n_y(n_p + 1) \times 1$ vectors Y and Y' which have been augmented by the sensitivity variables:

$$Y = \begin{bmatrix} y \\ s_1 \\ \vdots \\ s_{n_p} \end{bmatrix}, \quad Y' = \begin{bmatrix} y' \\ s_1' \\ \vdots \\ s_{n_p}' \end{bmatrix}.$$

The user must initialize these vectors so that both the original ODE/DAE and the sensitivity system are consistent. The parameters of the problem must be located somewhere within the user array RPAR, and the first n_p elements of IPAR must point to the locations of those parameters within RPAR as follows: IPAR(i) contains the location of the ith parameter in RPAR. For example, if IPAR(2) = 5, then RPAR(IPAR(2)) = RPAR(5) = value of parameter p_2.

The user can specify whether the sensitivity equations are to be evaluated analytically or approximated by finite differences. To evaluate them analytically, the user must provide a RES routine which produces the residual of the original DAE *and the sensitivity equations as well*.

DASPKSO utilizes the same sensitivity subroutine as DASSLSO. The extensions to the user interface are identical to those of DASSLSO with a few very minor exceptions.

In addition to DASSLSO and DASPKSO, a stand-alone routine (SENSD) which performs a sensitivity analysis of a derived quantity has been constructed. For example, one might want to know the sensitivity of a norm of the solution with respect to perturbations in the parameters. This routine approximates the analytic sensitivity equations by finite differencing the derived quantity $Q(t, y, y', p)$ ($p \in \mathbb{R}^{n_p}$, $y \in \mathbb{R}^{n_y}$, and $Q \in \mathbb{R}^{n_q}$), using

$$\frac{dQ(t,y,y',p)}{dp_i} = \frac{\partial Q}{\partial y}\frac{dy}{dp_i} + \frac{\partial Q}{\partial y'}\frac{dy'}{dp_i} + \frac{\partial Q}{\partial p_i}.$$

Expanding $Q(t, y, y', p)$ in a Taylor's series about y results in

$$Q(t, y + \delta_i s_i, y' + \delta_i s_i', p + \delta_i e_i)$$
$$= Q(t, y, y', p) + \delta_i \frac{\partial Q}{\partial y} s_i + \delta_i \frac{\partial Q}{\partial p_i} + \delta_i \frac{\partial Q}{\partial y'} s_i' + O(\delta_i^2)$$

so that

$$\frac{dQ(t,y,y',p)}{dp_i} \approx \frac{Q(t, y + \delta_i s_i, y' + \delta_i s_i', p + \delta_i e_i) - Q(t,y,y',p)}{\delta_i}.$$

This, of course, is one of many possible finite difference schemes which can be used. In the code, central differencing is an option as well. The routine SENSD is called after a successful return from a call to DASSLSO or DASPKSO and must be provided with a function which defines the derived quantity Q.

Finally, we note that a naive approach to sensitivity analysis which has often been used in practice is to perturb each parameter slightly, compute the numerical solution, and approximate the sensitivity by the difference between this perturbed solution and the unperturbed solution, divided by the perturbation. The above-described algorithms are much more reliable and efficient. The derivatives are computed with much less error because the solutions to be differenced are in effect computed simultaneously, with the same sequence of stepsizes and orders. The size of the increment is adjusted dynamically as the size of the solution and sensitivity changes, yielding better scaling for the increment and greater accuracy for the sensitivities. Significant economies in the linear algebra are achieved by exploiting the structure of the sensitivity system, and sensitivities of derived quantities are computed at virtually no additional cost.

Obtaining Consistent Initial Conditions in DASPK

When using either of the solvers DASSL or DASPK or the extensions for sensitivity analysis, the integration must be started with a consistent set

7.5. THE DASSL FAMILY

of initial conditions y_0 and y'_0. Consistency requires, in particular, that $F(t_0, y_0, y'_0) = 0$. Usually, not all of the components of y_0 and y'_0 are known directly from the original problem specification. The problem of finding consistent initial values can be a challenging task. The present DASSL and DASPK solvers offer an option for finding consistent y'_0 from a given initial y_0, by taking a small artificial step with the backward Euler method. However, initialization problems do not always arise in this form, and even for the intended problem type, that technique is not always successful. In any case, it is unsatisfactory in that it produces values at $t = t_0 + h$ (h = stepsize) rather than at $t = t_0$. Brown, Hindmarsh, and Petzold have recently proposed an alternative procedure for a class of index one DAE initialization problems [277]. This method, in combination with the modified Newton methods of DASSL or the Newton-Krylov methods of [278, 279], yields an algorithm which converges nearly as rapidly as the underlying Newton or Newton-Krylov method. The new method is very convenient for the user, because it makes use of the Jacobian or preconditioner matrices which are already required in DASSL or DASPK. The new initialization procedures replace the old option in the latest release of DASPK, which has recently become available. The methods can also be extended to higher index Hessenberg DAE systems; a paper on this is in progress.

The new initialization technique is applicable to two classes of index one initialization problems. Initialization Problem 1 is posed for systems of the form

$$\begin{aligned} f(t, u, v, u') &= 0, \\ g(t, u, v) &= 0, \end{aligned} \quad (7.5.28)$$

where $u, f \in \mathbf{R}^{N_d}$ and $v, g \in \mathbf{R}^{N_a}$, with the matrix $f_{u'} = \partial f / \partial u'$ square and nonsingular. The problem is to find the initial value v_0 of v when the initial value u_0 for u is specified. Actually, the procedure applies to a somewhat more general class of problems [277] but for the sake of clarity we will focus here on (7.5.28).

In Initialization Problem 2, which is applicable to the general index one system

$$F(t, y, y') = 0,$$

the initial derivatives are specified but all of the dependent variables are unknown. That is, we must solve for y_0 given y'_0. For example, beginning the DAE solution at a steady state corresponds to specifying $y'_0 = 0$.

The idea motivating the new software is to solve both of these initial condition problems with the help of mechanisms already in place for the solution of the DAE system itself, rather than requiring the user to perform a special computation for it. Consider first Initialization Problem 1 for the semi-explicit index one system (7.5.28), where $v_0 = v(t_0)$ is to be determined, given $u_0 = u(t_0)$ at the initial point $t = t_0$. We expand this problem to include the calculation of $u'_0 = u'(t_0)$. Thus we can form a nonlinear system in the N-vector

$$x = \begin{bmatrix} u'_0 \\ v_0 \end{bmatrix}, \qquad (7.5.29)$$

namely

$$F(x) \equiv \begin{bmatrix} f(t_0, u_0, v_0, u'_0) \\ g(t_0, u_0, v_0) \end{bmatrix} = 0 . \qquad (7.5.30)$$

A Newton iteration for the solution of $F(x) = 0$ would require the Jacobian matrix

$$F'(x) = \begin{bmatrix} f_{u'} & f_v \\ 0 & g_v \end{bmatrix} . \qquad (7.5.31)$$

By assumption, this matrix is nonsingular, at least in a neighborhood of the desired solution.

In the course of integrating a DAE system with DASSL or DASPK, the user must call upon one of several linear system algorithms to solve $N \times N$ linear systems at every time step. The linear systems have the form

$$A \Delta y = R ,$$

in which R is a residual vector, Δy is a correction to y, and the matrix A is the DAE system iteration matrix

$$A = \alpha \frac{\partial F}{\partial y'} + \frac{\partial F}{\partial y} . \qquad (7.5.32)$$

The user is encouraged to supply an approximation to A, for use either as the Newton matrix itself in the case of direct methods, or as a preconditioner in the case of a Krylov method. In the direct case, A is generated by difference quotient approximation if not supplied by the user. In the case of the system (7.5.28), we have

$$A = \alpha \begin{bmatrix} f_{u'} & 0 \\ 0 & 0 \end{bmatrix} + \begin{bmatrix} f_u & f_v \\ g_u & g_v \end{bmatrix} = \begin{bmatrix} \alpha f_{u'} + f_u & f_v \\ g_u & g_v \end{bmatrix} . \qquad (7.5.33)$$

In order to make use of A in solving $F(x) = 0$, we pick an artificial stepsize h, and set $\alpha = 1/h$ in (7.5.33). Then, to recover the block $f_{u'}$, we rescale the first block-column of A by h, using the scaling matrix

$$S = \begin{bmatrix} hI_d & 0 \\ 0 & I_a \end{bmatrix}, \qquad (7.5.34)$$

where I_d and I_a are the identity matrices of size N_d and N_a, respectively. Thus we consider the matrix

$$\bar{A}(x) = AS = \begin{bmatrix} f_{u'} + hf_u & f_v \\ hg_u & g_v \end{bmatrix}, \qquad (7.5.35)$$

7.5. THE DASSL FAMILY

evaluated at $t = t_0$, $(u, v) = (u_0, v_0)$, and $u' = u'_0$. Note that $\bar{A}(x) = F'(x) + hC(x)$ with

$$C(x) = \begin{bmatrix} f_u & 0 \\ g_u & 0 \end{bmatrix}.$$

Thus if h is small in some appropriate sense, we can expect that $\bar{A}$ will be a good approximation to $F'(x)$.

The new initialization procedure carries out a Newton-like iteration with corrections

$$\Delta x = -\bar{A}^{-1} F(x). \tag{7.5.36}$$

Each iteration calls on the linear system solution procedure that is to be used later in solving the DAE system itself. It also requires information about which components of y are differential and which are algebraic, in order to apply the correction Δx to the vectors y and y'. But otherwise, the procedure requires no additional information or methodology. Upon convergence, we have all of the components of $y(t_0)$ and we have the components of $y'(t_0)$ corresponding to u'_0, the derivatives of the differential variables. The remaining components of $y'(t_0)$, corresponding to v'_0, will simply be set to zero because the integration procedure is largely insensitive to these (since v' does not appear in (7.5.28)), and the first time step will produce accurate values for them.

For Initialization Problem 2, we are given the initial value of y' and must compute the initial y. In this case, we are simply interested in solving for $x = y_0$ in the system

$$F(t_0, x, y'_0) = 0, \tag{7.5.37}$$

with y'_0 given. We assume that this problem is well posed as provided by the user, so that $\partial F / \partial x$ is nonsingular in a neighborhood of the solution, including the initial guess supplied. As in the first problem, we will call for the user to supply the DAE iteration matrix A, but this time we set $\alpha = 0$, so that the matrix involved is simply $A = \partial F / \partial y$; there is no stepsize h. We then proceed with Newton iterations using A, with corrections

$$\Delta y_0 = -A^{-1} F(t_0, y_0, y'_0). \tag{7.5.38}$$

In order to improve the robustness of the Newton-type algorithm discussed above, the codes also employ a linesearch backtracking algorithm. Further details on the algorithms, including convergence results, implementation details, and numerical results, can be found in [277].

Availability

At the time of this writing, DASSL is available on the Internet via Netlib [307] or Xnetlib. To retrieve double precision DASSL from Netlib,

```
mail netlib@ornl.gov
Subject: send ddassl from ode
```

You will receive a message from Netlib with the DASSL source code. For single precision DASSL, replace ddassl with sdassl in the above message. The basic DASSL code is public domain software.

A root-finding version of DASSL called DASSLRT was discussed in Chapter 5. This code has been released and is now called DASRT. It also is available on the Internet via Netlib. Like DASSL, DASRT is public domain software.

The DASPK code for large-scale DAE systems is available by contacting the authors. Send email to Alan Hindmarsh,

```
mail na.hindmarsh@na-net.ornl.gov
```

and indicate whether you want the single or double precision version. Unlike DASSL and DASRT, the DASPK software is copyrighted. You will need to request permission of the authors to use it, which is usually granted for research purposes. A new version of DASPK incorporating the more robust initialization procedure described earlier in this section has recently been released.

Data-parallel (DASPKF90) and message-passing (DASPKMP) versions of DASPK which were written for the Thinking Machines CM-5 Computer are available [348]. Send email to Bob Maier,

```
mail maier@s1.arc.umn.edu
```

Jean-Phillipe Brunet [280] has modified DASPKF90 for inclusion in Thinking Machines Corporation's mathematical subroutine library.

At the time of this writing, the sensitivity software DASSLSO and DASPKSO are still under development, with release planned in late 1995. Contact Linda Petzold,

```
mail na.petzold@na-net.ornl.gov
```

for further details and an update on availability.

The latest versions of much of the DASSL family of codes will soon be available on the World Wide Web at the address

```
http://www.cs.umn.edu/~petzold
```

Potential users are encouraged to consult this source first.

Bibliography

[245] P. Amodio and F. Mazzia, *Boundary value methods for the solution of differential-algebraic equations*, Numer. Math., 66 (1994), 411-421.

[246] Th. Andrzejewki, H. G. Bock, E. Eich, and R. v. Schwerin, *Recent advances in the numerical integration of multibody systems*, in Advanced Multibody System Dynamics - Simulation and Software Tools, Kluwer Academic Publishers, Norwell, MA, 1993, 127-151.

[247] J. D. Aplevich, *Implicit Linear Systems*, Springer-Verlag, Berlin, 1991.

[248] C. Arévalo, *Matching the Structure of DAE's and Multistep Methods*, Ph. D. thesis, Department of Computer Science, Lund University, 1993.

[249] C. Arévalo and P. Lötstedt, *Improving the accuracy of BDF methods for index-3 differential algebraic equations*, BIT, to appear.

[250] C. Arévalo, C. Führer and G. Söderlind, *Stabilized multistep methods for index 2 Euler-Lagrange DAE's*, BIT, 1996, to appear.

[251] C. Arévalo and G. Söderlind, *Convergence of multistep discretizations of DAE's*, BIT, 35 (1995), 143-168.

[252] M. Arnold, K. Strehmel, and R. Weiner, *Half-explicit Runge-Kutta methods for semi-explicit differential equations of index 1*, Numer. Math., 64 (1993), 409-431.

[253] W. E. Arnoldi, *The principle of minimized iterations in the solution of the matrix eigenvalue problem*, Quart. J. Appl. Math., 9 (1951), 17-29.

[254] U. M. Ascher, H. Chin, and S. Reich, *Stabilization of DAE's and invariant manifolds*, Numer. Math., 67 (1994), 131-149.

[255] U. M. Ascher, H. Chin, L. R. Petzold, and S. Reich, *Stabilization of constrained mechanical systems with DAE's and invariant manifolds*, Mech. Structures Mach., 23 (1995), 135-158.

[256] U. Ascher and L. R. Petzold, *Projected implicit Runge-Kutta methods for differential-algebraic systems*, SIAM J. Numer. Anal., 28 (1991), 1097-1120.

[257] U. M. Ascher and L. R. Petzold, *The numerical solution of delay-differential algebraic equations of retarded and neutral type*, SIAM J. Numer. Anal., 32 (1995), 1635-1657.

[258] U. Ascher and L. R. Petzold, *Stability of computational methods for constrained dynamics systems*, SIAM J. Sci. Comput., 14 (1993), 95-120.

[259] U. Ascher and R. Spiteri, *Collocation software for boundary value differential-algebraic equations*, SIAM J. Sci. Comput., 15 (1994), 938-952.

[260] R. Bachman, L. Brüll, Th. Mrziglod, and U. Pallaske, *On methods of reducing the index of differential algebraic equations*, Comput. Chem. Engrg., 14 (1990), 1271-1273.

[261] Y. Bai, *A perturbed collocation method for boundary-value problems in differential algebraic equations*, J. Appl. Math. Comp., 45 (1991), 269-291.

[262] A. Barrlund, *Constrained least squares methods for linear time varying DAE systems*, Numer. Math., 60 (1991), 145-161.

[263] A. Barrlund, *Comparing stability properties of three methods in DAE's or ODE's with invariants*, BIT, 35 (1995), 1-18.

[264] A. Barrlund, *User guide to CLS – a Fortran code for the numerical solution of higher-index differential algebraic equation systems*, Department of Computing Science, Umea University, 1994.

[265] A. Barrlund and B. Kågström, *Analytical and numerical solutions to higher index linear variable coefficient DAE systems*, J. Comput. Appl. Math., 31 (1990), 305-330.

[266] L. T. Biegler, *Optimization strategies for complex process models*, Advances in Chemical Engineering, 18 (1992), 197-256.

[267] P. Bilardello, X. Joulia, J. M. LeLann, H. Delmas, and B. Koehret, *A general strategy for parameter estimation in differential-algebraic systems*, Comput. Chem. Engrg., 17 (1993), 517-525.

[268] C. Bischof, A. Carle, G. Corliss, A. Griewank, and P. Hovland, *ADIFOR - Generating derivative codes from Fortran programs*, Scientific Programming, 1 (1992), 11-29.

[269] W. Blajer, *Index of differential-algebraic equations governing the dynamics of constrained systems*, Appl. Math. Modelling, 16 (1992), 70-77.

[270] H. G. Bock and R. v. Schwerin, *An inverse dynamics ADAMS-method for constrained multibody systems*, Heidelberg, Preprint 93-27, 1993.

[271] V. Brasey and E. Hairer, *Half-explicit Runge-Kutta methods for differential-algebraic systems of index 2*, SIAM J. Numer. Anal., 30 (1993), 538-552.

[272] K. E. Brenan, *Differential-algebraic issues in the direct transcription of path constrained optimal control problems*, Annals of Numerical Mathematics, 1 (1994), 247-263.

[273] K. E. Brenan, *A large-scale generalized reduced gradient algorithm for flight optimization and sizing problems*, The Aerospace Corporation, ATR-94(8489)-3, September 1994.

[274] P. N. Brown and A. C. Hindmarsh, *Matrix-free methods for stiff systems of ODE's*, SIAM J. Numer. Anal., 23 (1986), 610-638.

[275] P. N. Brown and A. C. Hindmarsh, *Reduced storage matrix methods in stiff ODE systems*, J. Appl. Math. Comp., 31, (1989), 40-91.

[276] P. N. Brown, A. C. Hindmarsh, and L. R. Petzold, *Using Krylov methods in the solution of large-scale differential-algebraic systems*, SIAM J. Sci. Comput., 15 (1994), 1467-1488.

[277] P. N. Brown, A. C. Hindmarsh, and L. R. Petzold, *Consistent initial condition calculation for differential-algebraic systems*, Technical Report, Lawrence Livermore National Laboratory, 1995.

[278] P. N. Brown and Y. Saad, *Hybrid Krylov methods for nonlinear systems of equations*, SIAM J. Sci. Comput., 11 (1990), 450-481.

[279] P. N. Brown and Y. Saad, *Convergence theory of nonlinear Newton-Krylov algorithms*, SIAM J. Optim., 4 (1994), 297-330.

[280] J. P. Brunet, *CMDASPK User's Guide, Version 1.0*, Thinking Machines Corporation, 1994.

[281] P. Bujakiewicz, *Maximum weighted matching for high index differential algebraic equations*, Thesis, Technical University Delft, 1994.

[282] M. V. Bulatov, *Transformations of differential-algebraic systems of equations*, Comp. Maths. Math. Phys., 34 (1994), 301-311.

[283] J. C. Butcher, *The Numerical Analysis of Ordinary Differential Equations, Runge-Kutta and General Linear Methods*, John Wiley & Sons, New York, 1987.

[284] G. D. Byrne and W. E. Schiesser, editors, *Recent Developments in Numerical Methods and Software for ODE's/DAE's/PDE's*, World Scientific, Singapore, 1991.

[285] G. D. Byrne, *Pragmatic experiments with Krylov methods in the stiff ODE setting*, in Computational Ordinary Differential Equations, J. R. Cash and I. Gladwell, eds., Oxford University Press, Oxford, 1992, 323-356.

[286] S. L. Campbell, *Least squares completions for nonlinear differential algebraic equations*, Numer. Math., 65 (1993), 77-94.

[287] S. L. Campbell, *Numerical methods for unstructured higher index DAE's*, Annals of Numerical Mathematics, 1 (1994), 265-278.

[288] S. L. Campbell, *DAE approximations of PDE modeled control problems*, Proc. IEEE Mediterranean Symposium on New Directions in Control and Automation, Crete, 1994, 407-414.

[289] S. L. Campbell, *High index differential algebraic equations*, Mech. Structures Mach., 23 (1995), 199-222.

[290] S. L. Campbell, *Linearization of DAE's along trajectories*, Z. Angew. Math. Phys. (ZAMP), 46 (1995), 70-84.

[291] S. L. Campbell, *Nonregular 2D descriptor delay systems*, IMA J. Math. Control Appl., 12 (1995), 57-67.

[292] S. L. Campbell and C. W. Gear, *The index of general nonlinear DAE's*, Numer. Math., to appear.

[293] S. L. Campbell and E. Griepentrog, *Solvability of general differential algebraic equations*, SIAM J. Sci. Comput., 16 (1995), 257-270.

[294] S. L. Campbell and B. Leimkuhler, *Differentiation of constraints in differential algebraic equations*, J. Mech. Struct. Mach., 19 (1991), 19-40.

[295] S. L. Campbell and W. Marszalek, *ODE/DAE integrators and MOL problems*, Z. Agnew. Math. Mech., to appear.

[296] S. L. Campbell and E. Moore, *Progress on a general numerical method for nonlinear higher index DAE's II*, Circuits Systems Signal Process., 13 (1994), 123-138.

[297] S. L. Campbell and E. Moore, *Constraint preserving integrators for general nonlinear higher index DAE's*, Numer. Math., 69 (1995), 383-399.

[298] S. L. Campbell, E. Moore, and Y. Zhong, *Utilization of automatic differentiation in control algorithms*, IEEE Trans. Automat. Control, 39 (1994), 1047-1052.

BIBLIOGRAPHY 243

[299] S. L. Campbell, N. K. Nichols, and W. J. Terrell, *Duality, observability, and controllability for linear time varying descriptor systems*, Circuits Systems Signal Process., 10 (1991), 455-470.

[300] P. Chartier, *General linear methods for differential-algebraic equations of index one and two*, Preprint, INRIA, 1993.

[301] C. Chen and C. Hwang, *Convergence analysis of a computational method for optimal control of non-linear differential-algebraic systems*, Internat. J. Systems Sci., 21 (1990), 2337-2350.

[302] Y. Chung and W. Westerberg, *A proposed numerical algorithm for solving nonlinear index problems*, Ind. Engrg. Chem. Res., 29 (1990), 1234-1239.

[303] L. Dai, *Singular Control Problems*, Springer-Verlag, Berlin, New York, 1989.

[304] R. S. Dembo, S. C. Eisenstat, and T. Steihaug, *Inexact Newton methods*, SIAM J. Numer. Anal., 19 (1982), 400-408.

[305] J. Demmel and Bo Kagstrom, *The generalized Schur decomposition of an arbitrary pencil $A - \lambda B$: Robust software with error bounds and applications. Part I: Theory and algorithms*, ACM Trans. Math. Software, 19 (1993), 160-174.

[306] J. Demmel and Bo Kagstrom, *The generalized Schur decomposition of an arbitrary pencil $A - \lambda B$: Robust software with error bounds and applications. Part II: Software and applications*, ACM Trans. Math. Software, 19 (1993), 175-201.

[307] J. Dongarra and E. Grosse, *Distribution of mathematical software via electronic mail*, Comm. ACM, 30 (1987), 403-407.

[308] E. Eich, *Convergence results for a coordinate projection method applied to mechanical systems with algebraic constraints*, SIAM J. Numer. Anal., 30 (1993), 1467-1482.

[309] E. Eich, C. Führer, and J. Yen, *On the error control for multistep methods applied to ODE's with invariants and DAE's in multibody dynamics*, Mech. Structures Mach., 23 (1995), 159-180.

[310] M. Fliess, J. Lévine, and P. Rouchon, *A simplified approach of crane control via a generalized state-space model*, Proc. 30th IEEE Conference on Decision and Control, 1991, 736-741.

[311] M. Fliess, J. Lévine, and P. Rouchon, *Index of implicit time-varying linear differential equation: A noncommutative linear algebraic approach*, Linear Algebra Appl., 186 (1993), 59-71.

[312] C. W. Gear, *Differential algebraic equations, indices, and integral algebraic equations*, SIAM J. Numer. Anal., 27 (1990), 1527-1534.

[313] C. W. Gear and J. B. Keiper, *The analysis of generalized BDF methods applied to Hessenberg form DAE's*, SIAM J. Numer. Anal., 28 (1991), 833-858.

[314] A. H. W. (T) Geerts and J. M. Schumacher, *Impulsive-smooth behavior in multimode systems. Part I: State-space and polynomial representations*, Preprint, 1994.

[315] A. H. W. (T) Geerts and J. M. Schumacher, *Impulsive-smooth behavior in multimode systems. Part II: Minimality and equivalence*, Preprint, 1994.

[316] P. E. Gill, W. Murray, and M. H. Wright, *Practical Optimization*, Academic Press, New York, 1981.

[317] E. Griepentrog, *Index reduction methods for differential-algebraic equations*. Seminarberichte Nr. 92-1, Humboldt-Universität zu Berlin, Fachbereich Mathematik, 1992, 14-29.

[318] E. Griepentrog, M. Hanke, and R. März, *Toward a better understanding of differential algebraic equations (Introductory survey)*, Seminarberichte Nr. 92-1, Humboldt-Universität zu Berlin, Fachbereich Mathematik, 1992, 1-13.

[319] A. Griewank and G. F. Corliss, editors, *Automatic Differentiation of Algorithms: Theory, Implementation, and Application*, SIAM, Philadelphia, 1991.

[320] A. Griewank, D. Juedes, and J. Srinivasan, *ADOL-C: A package for the automatic differentiation of algorithms written in C/C++*, Preprint MCS-P180-1190, Mathematics and Computer Science Division, Argonne National Laboratory, 1991.

[321] E. Hairer and L. O. Jay, *Implicit Runge-Kutta methods for higher index differential-algebraic systems*, WSSIAA, Contributions in Numerical Mathematics, 2 (1993) 213-224.

[322] E. Hairer, S. Nörsett, and G. Wanner, *Solving Ordinary Differential Equations I: Nonstiff Problems, Second Revised Edition*, Springer-Verlag, Berlin, 1993.

[323] E. Hairer and G. Wanner, *Solving Ordinary Differential Equations II: Stiff and Differential-Algebraic Problems*, Springer-Verlag, Berlin, 1991.

[324] E. J. Haug and R. C. Deyo, editors, *Real-Time Integration Methods for Mechanical System Simulation*, Springer, Berlin, New York, 1991.

[325] D. J. Hill and I. M. Y. Mareels, *Stability theory for differential algebraic systems with applications to power systems*, IEEE Trans. Circuits Systems, 37 (1990), 1416-1423.

[326] M. Hou, Th. Schmidt, R. Schüpphaus, and P. C. Müller, *Normal form and Luenberger observer for linear mechanical descriptor systems*, J. Dynamic Systems Measurement Control, 115 (1993), 611-620.

[327] L. O. Jay, *Convergence of a class of Runge-Kutta methods for differential-algebraic systems of index 2*, BIT, 33 (1993), 137-150.

[328] L. O. Jay, *Symplectic partitioned Runge-Kutta methods for constrained Hamiltonian systems*, SIAM J. Numer. Anal., (1996), to appear.

[329] L. O. Jay, *Convergence of Runge-Kutta methods for differential-algebraic systems of index 3*, Appl. Numer. Math., 17 (1995), 97-118.

[330] L. O. Jay, *Runge-Kutta Type Methods for Index Three Differential-Algebraic Equations with Applications to Hamiltonian Systems*, Université de Genève, Département de Mathématiques, Genève 24, Switzerland, 1994.

[331] H. Krishnan and N. H. McClamroch, *Tracking in nonlinear differential-algebraic control systems with applications to constrained robot systems*, Automatica, 30 (1994), 1885-1887.

[332] A. Kröner, W. Marquardt, and E. D. Gilles, *Computing consistent initial conditions for differential-algebraic equations*, Computers & Chemical Engineering, 16 (Supplement) (1992), S131-S138.

[333] A. Kröner, W. Marquardt, and E. D. Gilles, *Computation of consistent initial conditions for differential-algebraic equations*, Numer. Math., 1994 (submitted).

[334] A. Kröner, W. Marquardt, and E. D. Gilles, *Steady states as consistent initial conditions for DAE's*, Preprint, 1994.

[335] A. Kumar and P. Daoutidis, *Feedback control of nonlinear differential-algebraic-equation systems*, AIChE J., 41 (1995), 619-636.

[336] P. Kunkel and V. Mehrmann, *A new class of discretization methods for the solution of linear differential-algebraic equations with variable coefficients*, SIAM J. Numer. Anal., to appear.

[337] P. Kunkel and V. Mehrmann, *Canonical forms for linear differential-algebraic equations with variable coefficients*, J. Comput. Appl. Math., 56 (1994), 225-251.

[338] P. Kunkel, V. Mehrmann, and W. Rath, *GELDA: A software package for the solution of general linear differential algebraic equations*, Technical Report SPC 95.8, Technische Universität Chemnitz-Zwickau, 1995.

[339] A. Kvaerno, *The order of Runge-Kutta methods applied to semi-explicit DAE's of index 1, using Newton-type iterations to compute the internal stage values*, University of Trondheim, Preprint, 1992.

[340] R. Lamour, *A shooting method for fully implicit index-2 differential-algebraic equations*, Preprint Nr. 94-13, Humboldt-Universität zu Berlin, Fachbereich Mathematik.

[341] S. L. Lee, *Krylov Methods for the Numerical Solution of Initial-Value Problems in Differential-Algebraic Equations*, Ph.D. Thesis, Computer Sci. Dept., Univ. Illinois, 1993.

[342] A. Lefkopoulos and M. A. Stadtherr, *Index analysis of unsteady-state chemical process systems- I: An algorithm for problem formulation*, Comput. Chem. Engrg., 17 (1993), 399-413.

[343] B. Leimkuhler and R. D. Skeel, *Symplectic numerical integrators in constrained Hamiltonian systems*, J. Comput. Phys., 112 (1994), 117-125.

[344] Ch. Lubich, *On projected Runge-Kutta methods for differential-algebraic equations*, BIT, 31 (1991), 545-550.

[345] Ch. Lubich, *Integration of stiff mechanical systems by Runge-Kutta methods*, Z. Agnew. Math. Phys., 44 (1993), 1022-1053.

[346] Ch. Lubich, *Extrapolation integrators for constrained multibody systems*, Impact Comput. Sci. Engrg., 3 (1991), 213-234.

[347] Ch. Lubich, U. Nowak, U. Pöhle, Ch. Engstler, *MEXX-Numerical software for the integration of constrained mechanical multibody systems*, Konrad-Zuse Zentrum für Information Stechnik, Berlin, 1993.

[348] R. S. Maier, L. R. Petzold and W. Rath, *Solving large-scale differential-algebraic equations via DASPK on the CM5*, Concurrency: Practice and Experience, 1995, to appear.

[349] C. Majer, W. Marquardt, and E. D. Gilles, *Reinitialization of DAE's after discontinuities*, Preprint, 1995.

[350] T. Maly and L. R. Petzold, *Numerical methods and software for sensitivity analysis of differential-algebraic systems*, Appl. Numer. Math., to appear.

[351] R. März, *Practical Lyapunov stability criteria for differential algebraic equations*, Preprint Nr. 91-28, Humboldt-Universität zu Berlin, Fachbereich Mathematik, Banach Center Publications, 1994.

[352] R. März, *On quasilinear index 2 differential algebraic equations*, Seminarberichte Nr. 92-1, Humboldt-Universität zu Berlin, Fachbereich Mathematik, 1992, 39-60.

[353] R. März, *Numerical methods for differential-algebraic equations*, Acta Numerica, (1992), 141-198.

[354] R. März, *Canonical projectors for differential algebraic equations*, Comput. Math. Appl., to appear.

[355] R. März and C. Tischendorf, *Solving more general index-2 differential algebraic equations*, Comput. Math. Appl., 28 (1994), 77-105.

[356] S. Mattsson and G. Söderlind, *Index reduction in differential-algebraic equations using dummy derivatives*, SIAM J. Sci. Comput., 14 (1993), 677-692.

[357] E. Moore, *Constraint Preserving Multistep Integrators for Differential Algebraic Equations*, Ph.D. Thesis, North Carolina State University, 1994.

[358] V. V. Murata and E. C. Biscaia, Jr., *Identification and reduction methods of the index of chemical engineering models: Implementation aspects using algebraic computation*, Preprint, 1995 (Short version to appear in Proc. ICIAM95).

[359] C. C. Pantelides and P. I. Barton, *Equation-oriented dynamic simulation current status and future perspectives*, European Symposium on Computer Aided Process Engineering-2, 1994, S263-S285,

[360] K. Park, P. Gill, L. Petzold, J. B. Rosen, and L. Jay, *Optimal numerical control of nonlinear parabolic partial differential equations*, in preparation.

[361] L. R. Petzold, *Numerical solution of differential-algebraic equations in mechanical systems simulation*, Physica D, 60 (1992), 269-279.

[362] L. R. Petzold and F. A. Potra, *ODAE methods for the numerical solution of Euler-Lagrange equations*, Appl. Numer. Math., 10 (1992), 397-413.

[363] L. R. Petzold, Y. Ren, and T. Maly, *Regularization of higher-index differential algebraic equations with rank-deficient constraints*, SIAM J. Sci. Comput., to appear.

[364] K. G. Pipilis, *High Order Moving Finite Element Methods for Systems Described by Partial Differential-Algebraic Equations*, Ph.D. Thesis, Dept. Chemical Engineering and Chem. Tech., Imperial College of Science, Technology, and Medicine, 1990.

[365] F. A. Potra, *Implementation of linear multistep methods for solving constrained equations of motion*, SIAM J. Numer. Anal., 30 (1993), 74-789.

[366] F. A. Potra, *Numerical methods for differential-algebraic equations with application to real-time simulation of mechanical systems*, Z. Angew. Math. Mech., 74 (1994), 177-187.

[367] F. Potra, *Runge-Kutta integrators for multibody dynamics*, Mech. Structures Mach., 23 (1995), 181-198.

[368] F. A. Potra and J. Yen, *Implicit numerical integration for Euler-Lagrange equations via tangent space parameterization*, Mech. Structures Mach., 19 (1991), 77-98.

[369] F. A. Potra and W. C. Rheinboldt, *Differential-geometric techniques for solving differential algebraic equations*, in *Real-Time Integration Methods for Mechanical System Simulation*, E. J. Haug and R. C. Deyo, eds., Computer and Systems Sciences 69, Springer-Verlag, Berlin, 1991, 155-191.

[370] P. J. Rabier and W. C. Rheinboldt, *A general existence and uniqueness theorem for implicit differential algebraic equations*, Differential Integral Equations, 4 (1991), 563-582.

[371] P. J. Rabier and W. C. Rheinboldt, *A geometric treatment of implicit differential-algebraic equations*, J. Differential Equations, 109 (1994), 110-146.

[372] P. J. Rabier and W. C. Rheinboldt, *On impasse points of quasilinear differential algebraic equations*, J. Math. Anal. Appl., 181 (1994), 429-454.

[373] P. J. Rabier and W. C. Rheinboldt, *On the computation of impasse points of quasilinear differential algebraic equations*, Math. Comp., 62 (1994), 133-154.

[374] P. J. Rabier and W. C. Rheinboldt, *Classical and generalized solutions of time-dependent linear differential algebraic equations*, Linear Algebra Appl., to appear.

[375] P. J. Rabier and W. C. Rheinboldt, *Time-dependent, linear DAE's with discontinuous inputs*, Linear Algebra Appl., to appear.

[376] P. J. Rabier and W. C. Rheinboldt, *Discontinuous solutions of semilinear differential- algebraic equations I: Distribution solutions*, Nonlinear Anal., to appear.

[377] P. J. Rabier and W. C. Rheinboldt, *Discontinuous solutions of semilinear differential-algebraic equations II: p-consistency*, Nonlinear Anal., to appear.

[378] P. J. Rabier and W. C. Rheinboldt, *On the numerical solution of the Euler-Lagrange equations*, SIAM J. Numer. Anal., 32 (1995), 318-329.

[379] H. Rabitz, M. Kramer, and D. Dacol, *Sensitivity analysis in chemical kinetics*, Ann. Rev. Phys. Chem. 34, 419-461 (1983).

[380] W. Rath, *Canonical forms for linear descriptor systems with variable coefficients*, Technical Report SPC 95.16, Technische Universität Chemnitz-Zwickau, 1995.

[381] S. Reich, *On a geometrical interpretation of differential-algebraic equations*, Circuits Systems Signal Process., 9 (1990), 369-382.

[382] S. Reich, *On the local qualitative behavior of differential-algebraic equations*, Circuits Systems Signal Process., to appear.

[383] S. Reich, *Symplectic integration of constrained Hamiltonian systems by composition methods*, Technical Report, 1994.

[384] K. J. Reinschke, *Graph-theoretic approach to symbolic analysis of linear descriptor systems*, Linear Algebra Appl., 197 (1994), 217-244.

[385] Y. Saad, *Krylov subspace methods for solving large unsymmetric linear systems*, Math. Comp., 37 (1981), 105-126.

[386] Y. Saad and M. H. Schultz, *GMRES: a generalized minimal residual algorithm for solving nonsymmetric linear systems*, SIAM J. Sci. Statist. Comput., 7 (1986), 856-869.

[387] J. M. Sanz-Serna, *Symplectic integrators for Hamiltonian problems: an overview*, Acta Numerica (1991), 243-286.

[388] B. Simeon, *MBSPACK - Numerical integration software for constrained mechanical motion*, Surv. on Math. for Industry, 1995, to appear.

[389] G. Söderlind, *Remarks on the stability of high-index DAE's with respect to parametric perturbations*, Computing, 49 (1992), 303-314.

[390] R. Spiteri, U. Ascher, and D. Pai, *Numerical solution of differential systems with algebraic inequalities arising in robot programming*, Proc. IEEE Conf. on Robotics & Automation, Nagoya, 1995.

[391] B. L. Stevens, *Derivation of aircraft linear state equations from implicit nonlinear equations*, Proc. 29th Conf. Decision & Control, 1990, 465-469.

[392] P. Tanartkit and L. T. Biegler, *Stable decomposition for dynamic optimization*, Industrial and Engineering Chemistry Research, 34 (1995), 1253-1266.

[393] C. Tischendorf, *Feasibility and stability behavior of the BDF applied to index-2 differential algebraic equations*, Z. Angew. Math. Mech., 12 (1995), 927-946.

[394] C. Tischendorf, *On stability of solutions of autonomous index-1 tractable and quasilinear index-2 tractable DAE's*, Circuits Systems Signal Process., 13 (1994), 139-154.

[395] J. Unger, A. Kröner, and W. Marquardt, *Structural analysis of differential-algebraic equation systems - theory and applications*, Comput. Chem. Engrg., 19 (1995), 867-882.

[396] P. E. van Keken, D. A. Yuen, and L. R. Petzold, *DASPK: A new high order and adaptive time-integration technique with applications to mantle convection with strongly temperature- and pressure-dependent rheology*, Geophys. Astrophys. Fluid Dynamics, to appear.

[397] S. Vasantharajan and L. T. Biegler, *Simultaneous strategies for optimization of differential-algebraic systems with enforcement of error criteria*, Comput. Chem. Engrg., 14 (1990), 1083-1100.

[398] V. S. Vassiliadis, R. W. H. Sargent, and C. C. Pantelides, *Solution of a class of multistage dynamic optimization problems 2. Problems with path constraints*, Ind. Engrg. Chem. Res., 33 (1994), 2123-2133.

[399] V. Venkatasubramanian, H. Schättler, and J. Zaborszky, *A stability theory of large differential algebraic systems: A taxonomy*, Systems Science and Mathematics Report SSM 9201, Washington University, St. Louis, 152 pages.

[400] V. Venkatasubramanian, H. Schättler, and J. Zaborszky, *Analysis of local bifurcation mechanisms in large differential-algebraic systems such as the power system*, Proc. 32. Conference on Decision and Control, 1993, 3727-3733.

[401] T. Yamada and D. G. Luenberger, *Algorithms to verify generic causality and controllability of descriptor systems*, IEEE Trans. Automat. Control, AC-30 (1985), 874-880.

[402] J. Yen, *Constrained equations of motion in multibody dynamics as ODE's on manifolds*, SIAM J. Numer. Anal., 30 (1993), 553-568.

[403] J. Yen and L. R. Petzold, *On the numerical solution of constrained multibody dynamic systems*, Preprint, Dept. of Computer Science, University of Minnesota, 1994.

[404] J. Yen and L. R. Petzold, *Convergence of the iterative methods for coordinate-splitting formulation in multibody dynamics*, Preprint, Dept. of Computer Science, University of Minnesota, 1995.

Index

A-stable Gauss-Legendre 99
A-stable IRK 76
absolute coordinates 152
acceleration-level constraint 153
adaptive mesh methods 148
adaptive moving mesh 12, 144
algebraic constraints 19, 114, 121
algebraic order of an IRK 82
algebraic variables 41, 64, 67, 157
analytic equivalence 30
analytically equivalent 25
angle of attack 157
approximation 31
asymptotic expansion 111
ATOL 124, 131

B-consistency 106
B-convergence 107
backward differentiation formulas 41
banded matrix 122, 131
bank angle 157
BDF 41, 63, 68, 144, 155
beam deflection 27
bifurcation 16
block implicit method 77, 83, 88
boundary conditions 135
boundary value problem 100
Butcher diagram 75

canonical singular system 79
cheap control 10
chemical engineering 137
chemical process control 157
chemical reactor 8, 35
choked flow 135
collocation method 88, 100
collocation SIRK 105

combustion modeling 11, 12, 108
completely singular system 79
computer-generated model 2
Concurrent DASSL 137
condition number 144
consistent initial conditions 19, 31, 130, 138
consistent of local order 69
consistent one-leg 71
constant coefficient order 85
constrained mechanical system 4, 58, 140, 150
constrained variational problem 35
constraint equations 36, 63, 136, 140, 151
continuity of solutions 31
contractivity condition 110
control variables 157
coordinate change 25, 33
corrector formula 119
crossrange capability 163

DAE 1
DASP3 129
DASSAC 137
DASSLRT 136
DASSL 42, 46, 54, 76, 103, 115, 146, 162, 166
defect correction 141
DELTA 129
derivative array 29, 32, 139
descriptor form 151
descriptor system 2, 6
diagnostics 116
diagonally-implicit IRK 77
differential variables 41, 64, 67, 157
differential-algebraic equation 1

253

DIRK 77, 98, 144
discontinuities 76, 108, 117, 133, 141
distributional limit 62
distributional solution 58
drag-acceleration constraint 164
drift 141, 156
dynamic order 38

electric power system 76
electrical circuit 6
error tolerances 135, 162
ERR 126
EST 128
Euler equations 150
Euler-Lagrange equations 4, 28, 37
explicit method 10
explicit ODE 1
extrapolation method 86, 88, 99, 105, 108, 115, 141

FACSIMILE 117
filtering 147
finite differencing 124
fixed coefficient 118
fixed leading coefficient 118
fluid flow 135
FNS 155
fully-implicit IRK 77
fully-implicit DAE 3

gas transmission network 76
generalized state space 2
global error 45, 90
global index 17
GMRES method 137

HARWELL 137
Hazony section 9
heat equation 10
Hessenberg form 26, 34, 56, 61, 105, 165, 169
higher index 18, 26, 29, 134

IDID 130
ill-conditioning 144
implementable 42

implicit Euler method 23, 33, 41, 43, 166
implicit method 10
implicit midpoint method 72, 85, 99
implicit ODE 1
implicit Runge-Kutta 75
implicit trapezoid method 83, 108
implicit system 2
inconsistent initial conditions 133
index of a matrix 19
index of a pencil 19
index of a variable 38
index one 34
index reduction 38, 141
index zero 17
index 7, 17, 22, 25, 30, 33, 39, 160
inertial coordinates 158
INFO 130
initial boundary layer 43
initial conditions 16, 65
initial state vector 166
initial stepsize 128
inner solution 59
instability 127
integration error tests 146
internal local truncation error 89
internal stage order 89, 99, 107
interpolation error 126
interpolation 121
invariants 8, 36
IPAR 130
IRES 129
IRK order reduction 89
IRK 75, 144, 169
iteration error 123
iteration matrix 130, 144
iteration termination 123
IWORK 132

Jacobian approximation 131
Jacobian calculation 132
JAC 130, 133

Kronecker canonical form 18

L-stable methods 77, 86

INDEX

L-stable SIRK 105
Lagrange multiplier 5, 140
least squares 139
LIMEX 108, 117
linear constant coefficient 3, 18
linear in the derivative 4, 65
linear multistep methods 41, 47, 63
linear time varying DAE 3, 22
linearly implicit DAE 4, 99, 116
linearly implicit method 111
LINPACK 122
LMM 41
Lobatto IIIA 83, 88
Lobatto IIIA 77
Lobatto IIIC 99
local error 45, 69, 71, 78, 81
local index 24, 25, 33
local stage derivatives 81
local truncation error 125, 147
lower triangular 21
LSODE 127
LSODI 42, 46, 54, 76, 116, 118, 166

massively parallel computers 137
matrix pencil 18
mechanical systems 148
METAN1 108
method of lines 10, 132, 135, 137
MOL 10, 26
multibody formalism 152
multiple solutions 168

Navier-Stokes equations 12, 145
NEQ 130
nilpotency 18
nilpotent pencil 19
nilpotent 21
noncanonic 2
noncausal 2
nonstiff order 99
normal form 1
norm 123, 131
nullity 31
nullspace 24, 65
numerically consistent 56

observation 7
ODE 1
one-leg method 42, 63
one-step methods 108, 115
optimal control 5, 35
order conditions 101
order reduction 106
order selection 127
order of a singular arc 17
outer solution 59

parameter variation 2
parasitics 9
partial differential equations 10, 148, 177
path constraints 157, 160
path control problem 104
PDE's 10
penalty functions 142, 154
pencil regularization 61
pendulum problem 5, 103, 142, 151, 154
physical constraints 37
pole 21
position-level constraint 152
positive solution 131
predictor polynomial 118
prescribed path control 7, 35
problem formulation 144

quadratic regulator problem 6, 10, 26

Radau IIA 100
range 113
real analytic 28
reduced order model 9
reduction procedure 20, 24, 27, 160
regular DAE 23
regular pencil 18
regularization 61, 107, 142
relative coordinates 152, 158
RES 129
rigid bodies 150
robotics 5, 7, 35, 152, 157

root finding 135
roundoff error 84, 124, 145
row scaling 146
RPAR 130
RTOL 124, 131, 156
RWORK 132

semi-explicit DAE 3, 26, 34, 39
semi-explicit IRK 77
semi-implicit midpoint method 108
semistate 2
semilinear systems 107
sensitivity analysis 136
singly implicit IRK 77
singular iteration matrix 134
singular perturbation problem 9, 58, 106
singular singular perturbation problem 61
singular system 2
SIRK22 169
SIRK23 169
SIRK 77, 98, 144
smoothly 1-full 29
solution jump 168
solution of DAE 16
solvable 16, 18, 22, 23, 30
space shuttle 7
space vehicles 157
sparse 137
sparsity 2, 140
SPRINT 117, 140
stability condition 84
stability constant 84
stability 45, 103, 141
stabilized index two formulation 153
stable multistep method 70
stable one-leg method 72
stage derivatives 75
standard canonical form 27, 34
state equations 157
state space form 151
state variable constraints 157
state variables 157
steep transients 148

stepsize selection 162
stepsizes 127
STEPS 155
stiff differential equation 10, 58
stiff ODE 106
stiffly accurate methods 88
stiffly accurate 77
strict stability condition 84
strictly stable IRK 79, 84
structural property 21
structurally nilpotent 21, 27

Taylor coefficient 29
TERKM1 126
TERKM2 126
TERKP1 126
TERK 126
three-stage SIRK 105
TOUT 128
TPPC 157
trajectory prescribed path control 7, 157
transferable 73
transformed SIRK 105
tree-configured systems 152
triangular chain 28, 35

uniform index one 48, 73, 90
upper triangular 21

v-i characteristic 6
variable coefficient 118
variable stepsize 45, 54, 56, 57, 68, 118
velocity-level constraint 152

YPRIME 129